THIS BOOK COULD FIX YOUR LIFE

Since 1956, **New Scientist** has established a world-beating reputation for exploring and uncovering the latest developments and discoveries in science and technology, placing them in context and exploring what they mean for the future. Each week through a variety of different channels, including print, online, social media and more, *New Scientist* reaches over 5 million highly engaged readers around the world.

Helen Thomson is a freelance writer and consultant with *New Scientist*. She has also written for the *Guardian*, the *New York Times, Nature* and the BBC, and has won various awards for her journalism. Her book *Unthinkable: An Extraordinary Journey Through the World's Strangest Brains* was *The Times* Book of the Year in 2018. Helen has a BSc in Neuroscience and an MSc in Science Communication. She lives in London.

Books by *New Scientist* include

How to be Human
The Origin of (almost) Everything
The Brain
How Long is Now?
The Universe Next Door
This Book Will Blow Your Mind
This Book Could Save Your Life

THIS BOOK COULD FIX YOUR LIFE

THE SCIENCE OF SELF HELP

HELEN THOMSON

NewScientist

First published in Great Britain in 2021 by John Murray (Publishers)
First published in the United States of America in 2021
by Nicholas Brealey Publishing
Imprints of John Murray Press
An Hachette UK company

1

Copyright © New Scientist 2021

The right of New Scientist to be identified as the
Author of the Work has been asserted by them in accordance
with the Copyright, Designs and Patents Act 1988.

This book is intended for information purposes only, and should not
be taken as individual medical advice. If you have medical concerns,
you should consult a medical practitioner. You should also consult
a medical practitioner before making changes to your diet or exercise
regime, especially if you have pre-existing health conditions.

A CIP catalogue record for this title is available from the British Library

UK Trade Paperback ISBN 978-1-529-31136-5
US Trade Paperback ISBN 978-1-529-34617-6
UK eBook ISBN 978-1-529-31142-6
US eBook ISBN 978-1-529-34616-9

Typeset in Bembo by Palimpsest Book Production Ltd, Falkirk, Stirlingshire

Printed and bound in Great Britain by Clays Ltd, Elcograf S.p.A.

John Murray policy is to use papers that are natural, renewable and
recyclable products and made from wood grown in sustainable forests.
The logging and manufacturing processes are expected to conform
to the environmental regulations of the country of origin.

John Murray (Publishers)
Carmelite House
50 Victoria Embankment
London EC4Y 0DZ
www.johnmurraypress.co.uk

Nicholas Brealey Publishing US
Hachette Book Group
Market Place Center, 53 State Street,
Boston, MA 02109 USA
www.nbuspublishing.com

CONTENTS

INTRODUCTION

I'M AWFULLY embarrassed. This really is a terrible place for us to start. But I have to be honest: I've always hated self-help books.

Let me explain. When I finished university and was desperate for advice on how to make the most of my future, 'cosmic ordering' had become a popular niche in this genre. It had a simple premise: like a spiritual Amazon Prime, you wish for what you want, put it out into the universe and wait for it to be delivered. Hot on its heels came *The Secret*, a book that promised to deliver the secrets of getting whatever you want using a mix of magical thinking and dubious quantum physics. It sold three million copies.

Neither provided me with satisfactory guidance. Almost two decades later and you'd think that our lives would have progressed from this kind of mumbo-jumbo. But I'm not so sure. Lately, it seems as though the loudest voices in the self-help sphere come from celebrities and Instagram influencers sharing the supposed secrets of their success, health and happiness, all the while endorsing their 'must have' products, of course. None of their advice appears to have much in the way of scientific backing. Even a lot of health coverage in newspapers is smeared by misleading headlines or cherry-picked studies that help sell copies.

I've lost count of the number of times I've read about some

novel philosophy that guarantees a happier life, a brilliant new exercise routine promising maximum results with minimal effort, or a brain-training app that helps you to improve your willpower, attention or creativity. Most times, I find myself mentally screaming, 'Has anyone actually checked this works?'

I guess a lot of us feel the same way. It's certainly a sentiment I hear often from my colleagues at *New Scientist* magazine, whose job often involves sifting through a lot of mumbo-jumbo searching for solid facts. In 2019, one of them, Graham Lawton, created *This Book Will Save Your Life*, which rounded up the most rigorous recent health research featured in the magazine, and converted it into actionable advice on all the big health questions – nutrition, weight loss, exercise and ageing. But that still left plenty of questions about the other parts of our life we'd like to fix, where the answers lie mainly (but not exclusively, as we shall find out) in our brains: things like our happiness, our bad habits, our friendships, our love life, our confidence, our atrocious memories.

That is where I, and this book, come in. Again, it's based on the best recent scientific research that I and my *New Scientist* colleagues have unearthed. This time, the aim is to provide you with a comprehensive and evidence-based guide to a smarter, happier and less stressful life.

When I agreed to write this book, I couldn't know that I would end up writing most of it during a global pandemic lockdown. At the time of writing, the future still looks increasingly uncertain. There's never been a better time to understand exactly how your brain works and how to use it to maximise the best aspects of your life, or help you cope when things go wrong.

And there are life fixes that do work. That is one of the main reasons why I find a lot of self-help advice so frustrating. I have a degree in neuroscience and have spent most of my working life as a health journalist, poring over scientific studies that seek to understand the brain and the body and how they work together to control our behaviour and direct our lives. I know that thousands of scientists – nutritionists, behavioural analysts, anatomists, neuroscientists, psychologists, geneticists – have collectively spent millions of hours investigating and testing ways that we can overcome our vices, improve our health, become more confident, boost our chances of success and, ultimately, live a better life.

I know, for instance, that the simple fact of knowing that willpower is not a limited resource, depleted by effort, makes you have more of it. I know that you're probably missing out on a simple exercise in your weekly routine that could reduce your risk of diabetes, heart disease and cancer. I know a way of revising for an exam that could boost your result by 10 per cent. And I know that money isn't going to buy you happiness, but a walk through the woods just might.

Somewhere along the line, then, I have become something of an accidental self-help expert. And I try to practise what I preach. The first change I consciously made to improve my life was to practise mindfulness regularly. It cropped up so many times that it was hard to ignore – there's an expanding pile of evidence from many scientific disciplines of its benefits, from improving your immune system to reducing the risk of depression and helping to tackle chronic pain and stress.

Then came more tweaks here and there. A while back, I was lucky to be accepted on a medical 'bootcamp' at Harvard University, where over a few intensive days journalists get to hear about a range of cutting-edge research directly from the scientists involved. One of the researchers I met there had just discovered how blue light – like the kind that our phones emit – triggers parts of the brain responsible for alertness, and can interfere with the onset of sleep. I vowed never to take my phone to bed with me again. For a while I managed it, noticing definite benefits to my sleep (you can find out how else you can improve your sleep in Chapter 7).

Old habits are hard to break though, and soon my phone was back tucked under the covers. Like anyone, I find it easy to slip into bad habits, but yet again science can help. Learning how our brain forms habits (Chapter 8) turns out to be helpful for anyone trying to break them, not just those who have the most damaging addictions.

There are even regular moments in my life when I use the power of positive thinking to boost my confidence, change my emotions or exercise more productively, having learned about the ways in which reframing our thoughts can have an impact on our actions. Maybe I shouldn't have been quite so quick to shun *The Secret* after all.

I am not claiming to live the perfect life – and indeed, attempting perfection can be very damaging, as you'll find out in the final chapter of the book. And don't get me wrong, I'm still cautious when it comes to offering advice, even when it's evidence-based. I don't believe that there is a one-size-fits-all guide to being a better, smarter, fitter, richer, happier, more successful person. Even the cleverest scientists fall prey to the same mental biases that trip all of us up (find out how to overcome yours in Chapter 10). Science itself is nuanced, imperfect and incomplete. Sometimes new evidence contradicts the old. Sometimes (whisper it) it turns out to be completely wrong.

So what I won't be giving you in this book is a promise to

make you happy overnight, or reveal the secret to becoming the world's greatest lover, or give you a step-by-step plan to becoming the CEO of a major global corporation. You can turn to magazines, autobiographies and the thousands of blogs on the internet for that if you wish (I will tell you, however, why it's a statistical fallacy to attempt to pick out the ingredients that lead to any one person's success – see Chapter 11).

But I can offer some advice on the four aspects of your life that you can work on to help you increase your happiness, as I do in Chapter 2. And while I can't promise to improve your sex life, in Chapter 5 I will show you what some of the world's best psychologists and neuroscientists know about boosting your confidence and charisma, and what psychotherapists have discovered about holding on to the one you love. And I can share some ways to tweak the behaviours that can get you a foot in the door at work, overcome some of the unconscious biases that may be working against you, and ensure you're more persuasive when it comes to getting that pay rise (Chapter 11).

This book will also arm you with the tools you need to control your bad habits, like overeating, biting your nails or drinking, and explore what techniques will help you quit the worst ones, like smoking. It will reveal the secrets to a good night's sleep, how to exercise more productively, how to combat stress, how to avoid making stupid decisions, how not to get lost, how to avoid procrastinating and a whole lot more.

Any advice is only as good as the current scientific evidence. When advice is based on lots of gold-standard trials, I'll let you know. When it's new evidence or a preliminary hypothesis, I'll tell you. Where there's controversy, I'll do my best to give you both sides of the story. I'm here to hold your hand as we wade together through the slushy bog of self-help and point you towards a path that has the best evidence for actually, well, helping.

One more thing to note. I have tried to make this book as relevant to your life today as possible. So while there are some really exciting cutting-edge technologies being developed that you might have heard of, if they're not accessible to you right now, you won't find them in this book. Deep brain stimulation, for instance, is proving itself as a fantastic treatment for things like Parkinson's disease and depression, but it isn't something that's easily available outside a clinical trial. Likewise, neural implants have the potential to boost your memory and treat dementia, but it's unlikely that you'll be offered the chance to have an electronic circuit implanted in your brain any time soon.

Lastly, I think it's important not just to give 'rules' to live by, but to understand the reasoning behind those rules. That is why each chapter will delve into the studies and history behind each piece of advice. My aim here is not to tell you how to live your life. It is to help you navigate the noise and often overwhelming onslaught of contradictory and confusing messages that we read and hear. My goal is to show you the science, and let you use it how you will.

Without needing to ask the universe for a thing.

This book is intended for information purposes only and should not be taken as individual medical advice. If you have medical concerns you should consult a medical practitioner. You should also consult a medical practitioner before making changes to your diet or exercise regime, especially if you have pre-existing health conditions.

1

HOW NOT TO WORRY

IT IS A HOT and sticky June 2020 as I write this, and I am sitting in my tiny box room. It is a nursery masquerading as an office as my family enters month four of lockdown, amid the worldwide coronavirus pandemic. Starting at 6.30 a.m., my husband and I split each day in half – one of us cares for our daughter while the other goes to work upstairs. At 1 p.m., we swap shifts. At 6.30 p.m. we share an evening routine of bath, stories and bed. A quick dinner, half an hour of TV, and then bed for me, too. On repeat. Like many others, we've not seen our family or friends since Christmas. We spend only an hour or so a day in a nearby park, and mostly get our food via deliveries. Oh, and I'm also eight months pregnant. To say that this is a challenging time is an understatement.

Like most people during the coronavirus crisis, the worry has at times got to me. I worry about my daughter and how she's missing her pals at nursery, and fear I am only inadequately fulfilling her needs. I worry about whether I'm going to get this book finished on time, or whether I'll end up doing less than the perfect job I want to. I worry about how my elderly relatives and some of my single friends are doing during lockdown.

And then there's the uncertainty. By the time you read this, it will probably be next year or later, and who knows what might have happened between now and then? I don't know what the world is going to look like when my imminent arrival is old enough to say his first words or take his first steps. Will he be able to meet his immediate family then? Will the economy be back on its feet? Will I have a job to return to?

I suspect this story, or at least elements of it, will resonate with you. But although the uncertainties of the pandemic may have been of seemingly unparalleled dimension, the strange, extended state of limbo really just amplified what assails us all the time. Stress, uncertainty and anxiety are to a greater or lesser extent human universals. It's unsurprising, therefore, that many psychologists, neuroscientists, evolutionary biologists and other researchers have set out to understand their roots and so perhaps develop strategies to combat them.

That fact, plus recent events, make stress an ideal point to start off. Like many of the things we'll encounter in this book, our stress response, and its impact on our life, varies greatly. But our resilience to stress is also something we can learn to improve – if we have the right knowledge.

HOW TO UNDERSTAND STRESS

We all know a person like it: someone entirely unfazed by all that modern life throws at them. The one who uses a delayed train as an excuse to get stuck into a good book. The one who can make a joke ten seconds after breaking their ankle. The one who loves giving presentations and never falters under pressure.

Most of us are not like that. In 2018, the largest-known study of stress levels in the UK showed that three-quarters of people had been so stressed in the past year that they had felt over-whelmed or unable to cope – and my bet is those numbers haven't improved given all that's happened since. Stress can be so damaging to our well-being that, according to that same study, one in three of those people had been left feeling suicidal, and one in six had self-harmed.

The good news, however, is that by understanding how stress works – why it exists, its physiology, its good and bad sides – we can put ourselves on the road to becoming more like the unflappable people we might aspire to be.

Stress is caused by a series of events that start in the brain's amygdala, its evolved threat-detection module. The amygdala is linked to parts of the cortex near the front of the brain that process social information and help us make decisions. It combines information from the world with memories of similar situations and sends out two kinds of distress signals when it senses danger.

The first triggers our ancient fight-or-flight system. This pumps out adrenaline, making us more alert, breathing more efficiently and forcing blood into our muscles to keep them strong – the better to outrun that lion on the plains of Africa. The second prompts the release of hormones like cortisol, which keeps our stress response active and releases stored glucose to give us energy. It also suppresses our digestion and immune system to focus resources on the threat, and triggers inflammation. This causes swelling so that white blood cells can flood damaged tissue and engulf invading pathogens or help protect against injury. When the threat has passed, cortisol levels drop and our bodies return to their normal state.

While these mechanisms are universal, there are certain things that influence our reaction to stress. Genetics is one. Animal experiments suggest that genes involved in the production of a chemical called neuropeptide Y, or NPY, act as a kind of on–off switch for our entire stress response. When we face a threat, production soars, helping to trigger a rapid response, but levels quickly return to normal once the danger is over.

We know from studying special forces soldiers that they can sustain higher levels of NPY for longer than regular soldiers. Their NPY levels also return to baseline more quickly after an interrogation or sleep and food deprivation, showing they're better able to recover from stress. What's more, the more NPY a special forces soldier releases, the less confusion and fewer mental health issues they report during training.[1]

We all inherit variations of NPY genes, some of which protect against stress and others that increase our risk of an impaired stress response. We don't yet know whether special forces soldiers are more likely to have the protective variation, which makes them more likely to become an elite soldier in the first place, or whether their training gives them the enhanced NPY response.

Upbringing is another factor in how we deal with stress. Animal experiments demonstrate that early trauma affects an individual's reaction to stress as an adult. Testing this link in humans is difficult, given the diversity and severity of events that can have a negative impact on a child's development. But one unique study of children raised in Romanian orphanages, the Bucharest Early Intervention Project, provides clues that the same thing happens in humans.

The children in this study were either kept in care or fostered between the ages of six months and thirty months. Just before they reached their teens, the children's stress response was tested. Those who had been fostered had similar cortisol levels to a control group of children living in nearby families, but only if they had been placed in foster care before the age of two. Those fostered later or who remained in orphanages had a blunted stress response, producing less cortisol, which reflects underlying damage to the normal stress response. This blunted response is associated with long-term behavioural problems and an increased risk of depression.

The first two years of our life appear to be an extremely sensitive period in which our environment is particularly likely to cause changes to the brain that influence the stress response. The mechanisms are probably numerous, but one theory is that trauma can trigger an increase in myelin in the brain's grey matter, where the cell bodies of neurons are found. Myelin normally forms an insulating sheath around nerve fibres, but in grey matter it prevents new connections forming between neurons, and is linked with PTSD and depression.[2] If trauma happens at such a critical age, when the brain is particularly malleable, it's likely to cause the most devastating effects.

All of this suggests that we might be helpless in mediating our reaction to stress – but that's far from the case, as we'll find out now.

HOW TO BUILD RESILIENCE

In a modern context, our evolved systems that react to threats come in very handy. They allow you to jump out of the way of an oncoming vehicle, for instance, before you've even registered what's going on. They focus your thoughts in front of a crowded auditorium. They protect against any passing viruses or injury. But sometimes, modern life triggers our stress response unnecessarily. Instead of being confined to a specific event or damaged tissue, it keeps the entire body in a perpetual state of readiness for a threat that never arrives.

Low-level chronic stress creates a slew of health problems. With no let-up, raised levels of adrenaline and inflammation can damage blood vessels and increase the risk of heart attack and stroke. High levels of cortisol can cause digestive problems, weight gain and diabetes. And constant modulation of the immune system can lead to fatigue and mental health issues.

We are faced with a double-edged sword. We want to react to stress quickly to protect ourselves from danger, but we don't want to turn that into a long-lasting response. The simple solution is to de-stress. How quickly you can de-stress is called resilience. And there's plenty we can do, because not even genetics is destiny when it comes to building resilience. We are not stuck with the levels of NPY, the chemical that triggers our stress response, that our genes dictate. Meditation can help here. After a surprise simulated ambush, the NPY levels of US Marines who had recently completed an eight-week course on mindfulness returned to normal far quicker than those of soldiers without such training.[3]

As far as nurture is concerned, a loving upbringing, with plenty of social interaction, stimulation and a strong, supportive, dependable relationship with your primary caregivers, is perhaps the single most important factor in developing stress resilience. But adults who did not have a good childhood experience are by no means doomed. If you are able to find the same ingredients – dependable, supportive, close relationships – the ill-effects of childhood adversity can be reduced. Regardless of our upbringing, a number of studies suggest that cultivating your social networks is one of the best ways to build resilience against the tough moments that life throws at you. Put the work in, in advance, to make sure your social ties are healthy so that when stress occurs you can rely on them.

There are other simple things you can do to protect yourself from the physical and mental effects of stress. The role of diet in promoting happiness is an important one. Chemicals called resolvins, which our bodies synthesise from omega-3 fatty acids, help inflammation return to normal after the initial reaction is over. So make sure you get the recommended three weekly portions of oily fish, and plenty of seeds, kiwi fruit and leafy greens, which might help your body to reduce excessive inflammation and de-stress, while also decreasing the associated risk of heart disease.

Despite imparting physical stress on the body, exercise is also a great way of minimising mental stress. It works by reducing stress hormones such as cortisol and adrenaline and stimulating the release of endorphins. These feel-good chemicals are the body's natural painkillers and are part of the reason that exercise has been shown in numerous clinical trials to reduce depression and elevate mood – an effect known as 'runner's high'.

If running is too much work, try a bit of comedy. Laughing releases some of the same hormones as exercise and makes us less likely to ruminate on or re-experience stressful events. It also helps us build relationships, providing the social support that is key to resilience.

Regularly listening to music also seems to be beneficial, lowering blood glucose and making challenges seem less stressful. Listening to music in the presence of others may strengthen its stress-reducing effects, probably because the added social aspect heightens our sense of emotional well-being. There's also evidence that people who synchronise their movement, or sing and make music together can also decrease stress, through social bonding and the release of those wonderful feel-good hormones in the brain.[4]

But don't forget that a little stress isn't a bad thing. What doesn't kill you really does makes you stronger – a term scientists like to call 'hormesis'. Exposing animals to low levels of stress extends their longevity and seems to moderate their response to stress over the long term. For instance, a team led by Daniela Kaufer at the University of California, Berkeley, exposed rats to a stressful environment for three hours while tracking the development of new neurons in their hippocampus, a brain region responsible for memory. Intriguingly, these cells proliferated more in the stressed animals than in a control group. But the real surprise was the long-term effect. Rats that had been stressed did better in cognitive tests, even weeks later, specifically engaging their new neurons to help with these tasks.

There's some evidence of similar effects in humans, at least on general mortality. For instance, an analysis of 28,000 nuclear shipyard workers who had been exposed to low doses of radiation, an environmental stressor, found that their mortality rate was 24 per cent lower than workers who were not exposed to radiation. Similar findings are found in radiologists compared with other doctors.

Stressors like radiation kick-start the body's natural repair mechanisms. When cells are stressed, from heat or toxins or inflammation, protective proteins shield others from attack. Stress also triggers the production of a protein called sirtuin 1, which triggers the creation of antioxidants – nutrients that protect the brain from something called oxidative stress, which is a build-up of material that can damage brain cells.[5] In the face of light stress, these repair systems overcompensate, repairing unrelated damage and rejuvenating your cells.

I'm not suggesting booking an extra X-ray, because there are probably better ways to expose yourself to small amounts of stress. Anecdotal reports from regular fasters say they believe it makes them able to think more clearly. The effect hasn't yet been confirmed in lab tests, but it makes evolutionary sense: if you're deprived of food you need a clear head to help you find some. Modern hunter-gatherers in Papua New Guinea also prefer to hunt on an empty stomach because they say it makes them sharper.

So what is the perfect amount of stress? Kaufer says she is often asked that question, but it is impossible to give an exact figure. The ideal amount is going to be different from one person to another. Something that makes us stressed one day might feel less daunting another. But if there is such a thing as beneficial stress, she says, then it's likely to be something you can pinpoint for yourself. It's probably not the stress that paralyses you, but the stress that you can push through, that makes you feel really great afterwards.

HOW TO DEAL WITH UNCERTAINTY

For me, and I think for a lot of people, much of the stress of the coronavirus lockdown was rooted in its uncertainty. Not for nothing is Limbo the first circle of hell in Dante's *Divine Comedy*. It is described as a place where people have no hope yet continue their longing, and what are first mistaken for cries of anguish are in fact sighs of sadness. Yes, that sounds about right.

Let's face it: not knowing isn't nice. We are curious. We like to know what is going on, what might happen and what the long-term effects of our actions might be. Our brains are geared towards predicting the future; our perception of the world is generated by combining memories of our past with information from our senses to make an educated guess about what is about to happen. Uncertainty can make us feel very uncomfortable.

Indeed, we find uncertainty so unsettling that some people would rather know they are going to receive an electric shock than wait for the possibility of one. This was shown when researchers at University College London got people to play a computer game where snakes were hidden behind certain rocks. Each time participants found a snake, they got a small electric shock. The computer measured uncertainty using the players' guesses and their stress response based on how much they sweated and their pupil size. People were more stressed if they were uncertain whether a shock was coming than if they knew they were definitely going to get zapped.

If we find ourselves in a situation where something bad is definitely going to happen, we can start to think about ways of coping. If we don't know whether a situation is going to be positive or negative, we remain in a state of mind less prepared for either outcome. That's the kind of paralysis that the coronavirus induced in a lot of us. It's also a feature of some of the economic fallout. In times of recession, it can in some ways be more stressful waiting for the possibility of a lay-off rather than just being told: 'You're fired.'

Where we sit on a spectrum of 'intolerance of uncertainty' affects how we experience not just big, all-embracing situations, but also smaller everyday scenarios, from waiting for a bus to waiting for an exam grade, the outcome of a difficult conversation with our boss, or news of a loved one in hospital.

When there's an ambiguous situation in our lives, you and I can have the exact same information and react in two completely different ways. Say your partner should have been back from work twenty minutes ago. Those with a low intolerance of uncertainty will assume they are stuck in traffic. A person with a high intolerance of uncertainty might immediately think they have been involved in an accident and worry until they arrive home.

Admittedly, sometimes a high intolerance of uncertainty is a good thing – you don't want an air traffic controller or brain surgeon to say, 'Well I don't know what's going to happen, but that's OK.' But for the most part, an extreme dislike of the unknown is undesirable. If you are someone who is less tolerant of uncertainty, you'll probably engage in several 'safety behaviours'. These are strategies that prevent undesirable outcomes in the future: phoning your partner all the time to check in with them is a prime example. While some safety behaviours allow us to minimise uncertainty, many, paradoxically, just make things worse. Safety behaviours in the absence of a realistic threat are actually maladaptive.

Here's a topical example. In a lab experiment, healthy people were told to engage in daily safety behaviours to prevent the spread of germs – washing their hands every time they touch a door handle, for instance. At the end of a week, they showed increased avoidance in contamination-related tests, and overestimated contamination threats. Too many safety behaviours mean that we never learn that uncertainty isn't always dangerous, and if you never have to experience negative outcomes, you never realise how good you might actually be at coping with them. Given this lab experiment has just been played out in real life around the globe, it's possible that we have all increased our intolerance to uncertainty in this particular aspect of life – the outcome of which is, ironically, uncertain.

So how can you work out how well you deal with uncertainty? You could use a scale developed by Michel Dugas at Concordia University in Montreal, Canada, in which you decide to what extent you agree with twenty-seven statements such as 'a small unforeseen event can ruin everything' or 'the smallest doubt can stop me from acting'.

Another technique that therapists use is called a 'catastrophising interview', in which you are asked to consider a current worry, such as the outcome of a job application. They then ask you what worries you about this situation. You might say you need the extra money. They would then ask you what worries you about that. 'What if I can't pay my rent?' you say. They ask you what worries you about that. 'Where would I borrow the money from? What if I default on my credit card? Would my children have to move schools?'

You examine the details of your worry until you get to the bottom of it, noting how many 'what if' scenarios you generate, how many future negative outcomes you imagine. The higher the number, the higher your intolerance of uncertainty. Normally, this test is used before and after a course of therapy to see whether it is working, rather than to understand how uncertainty might affect your life.

It's hard to compare our tolerance to uncertainty – like any trait it interacts with our life experience. If you're intolerant of uncertainty but your life is extremely predictable, you won't have any problem. If your life is chaotic, you might experience severe anxiety from the same level of intolerance. But don't underestimate the implications of all this.

Let's start with your health. Your doctor's personal uncertainty threshold has a huge impact on you. For instance, women are more likely to end up with a vaginal birth after a previous caesarean section if their doctor has a low intolerance of uncertainty. This trait also makes doctors more likely to offer a new genetic test, prescribe generic drugs, adopt a cutting-edge therapy and feel comfortable talking to patients about grief and loss.

However, doctors who are more intolerant of uncertainty are more likely to recommend a pregnancy termination following abnormal results from prenatal genetic tests, and are less willing to use newer therapies, such as cognitive behavioural therapy (CBT), say, for eating disorders. Doctors may also give different advice depending on how well they think their patients can cope with uncertainty.

Your own intolerance of uncertainty can also affect health outcomes. For instance, people who have a high tolerance of uncertainty have better emotional well-being after a cancer diagnosis and experience less distress after receiving genetic test results. It is also associated with a better quality of life and less irritability in people with epilepsy, as well as lower language impairment and fewer motor symptoms in those with Parkinson's disease. On the other hand, a high intolerance may make people more likely to adhere to their medication.

It also affects people's ability to cope with particular treatment regimes. For instance, men with localised prostate cancer may be able to choose a 'watch and wait' approach, whereby they have regular scans rather than immediate treatment that can have side effects including incontinence and impotence. This approach means enduring long periods of limbo between scans. It can be one of the most difficult choices a person ever has to make – and one that can causes rifts with loved ones, whose ability to cope with uncertainty differs from their own.

Contending with the unknown can place great strain on relationships. When couples with a high intolerance of uncertainty have difficulties they are more prone to leave each other immediately rather than wait and see what might happen. Or some people might have difficulty developing relationships in the first place, because they aren't prepared to go through that initial period of uncertainty – will they call? do they like me? should I ask them out to lunch? It can make you hesitant to form relationships, and when you do make the effort, you want too much certainty about the future from the start, which scares the other person off.

Merely understanding our own intolerance of uncertainty, and how it may differ from that of people around us, is a good start in working against some of the bad outcomes. But the good news is that our discomfort with the unknown can be manipulated. For instance, studies show that merely reading about a fictional character's high tolerance of uncertainty and putting yourself in that person's mindset can help you generate less stress in a later interview about real worries.

Another way around your fear of the unknown is to treat it as you would a phobia. If you're scared of dogs, a psychiatrist might expose you to the animal slowly to help you develop the understanding that most dogs are not dangerous. To cure fear of uncertainty you can experiment similarly with your safety behaviours. Say you're worried about letting your son go out alone. You make him call as soon as he leaves the house, get him to text on the bus, and call again once he's at his friend's house. In this scenario, you'd start by getting your son to call you just when he's on the bus, and then when he arrives. The next time he is out alone he just calls when he arrives, then on subsequent occasions does not have to call at all. It's about putting yourself in a situation where you can learn that uncertainty isn't dangerous. Studies show this leads to a decrease in general anxiety over time.

Minimising your safety behaviours without outside intervention is not impossible, but is difficult. Safety behaviours are really sneaky, they're hard to identify and we mostly don't realise we're doing them. But we are, increasingly so. A recent analysis of fifty-two studies of students shows that intolerance went up by about a fifth between 1994 and 2014. Mobile phones and internet access, which both grew rapidly over the same period, might be to blame – increasing safety behaviours by offering us immediate access to emergency services, loved ones and information isn't always helpful.

There are some strategies to help you cope with uncertainty that don't involve ditching your phone or resorting to professional help, however. For instance, throwing yourself into an engrossing task can provide a welcome distraction and make time pass more quickly. Practising mindfulness meditation, something we'll talk about in the next chapter, can help keep you in the moment, stopping you from agonising about future outcomes.

Don't forget that a degree of intolerance can be useful. Bracing for the worst can minimise the impact if bad news arises, although timing is everything. To avoid unnecessary worry, studies suggest that you need to assume the best for as long as possible before bracing for the worst towards the end of the wait.

It may also be helpful to concentrate on finding the silver lining in any potential bad news. In the 2016 US presidential election, supporters of Hillary Clinton who pre-emptively looked for the good in Donald Trump being elected were less shattered when he won. But be cautious, this strategy can backfire: Trump supporters who tried to find an upside to Clinton winning were less thrilled when their candidate did.

One final piece of advice for dealing with uncertainty: instead of weighing yourself down with worry or trying to problem-solve every eventuality, try sitting with that uncertainty for a while. You'll see that, most of the time, nothing particularly bad happens. And talk to others about how they cope. Taking a step back and realising that your way of dealing with uncertainty isn't set in stone, that others might not feel the same way about that same situation, and that your perspective is changeable, is one of the strongest messages you can take with you.

There may even be a silver lining to all the uncertainty we've just experienced over the past year or so. You might assume that there are more people having to deal with more uncertainty and more anxiety as a result of coronavirus and its legacy. But some researchers have suggested it could go the other way, making people less anxious, because they go on with their life even though they are experiencing more uncertainty. Future studies may show that successfully making it through something as big as Covid-19, a year-long limbo for some, can make the small things much easier to cope with.

HOW TO TREAT CHRONIC ANXIETY

As we've seen, stress is a natural response which puts our bodies in a state of preparedness, making us more aware of danger or uncertainty – something with an upside and a downside. In some people, however, the downside is emphasised – when anxious feelings just won't switch off. Stress turns into an anxiety disorder, something about 1 in 6 of us will contend with at some stage in our lives.

The most common type of anxiety disorder is social anxiety disorder, where you might believe, say, that blushing will result in people laughing at you, or shunning you, stopping you from going out in public. Another common type of anxiety disorder is panic disorder, characterised by recurring panic attacks that can cause dizziness or make breathing difficult. Fear of panic attacks can itself become a source of anxiety. Meanwhile gener- alised anxiety disorder is characterised by chronic worrying about a range of different things for at least six months.

Central to all this is the amygdala. During bouts of everyday stress and anxiety, this area of the brain switches on and then off again. However, for those with greater feelings of anxiety, it seems to get stuck in the on position. It amplifies negative information in your surroundings to make sure you pay attention to it, triggering your fight-or-flight response. It also slows digestion and makes us more susceptible to infection and pain.

A lot of this reaction is genetically mediated, and some of us are therefore naturally more prone to anxiety. Age and sex are factors. Women are twice as likely to develop anxiety disorders as men. In part this might be down to hormones and their influence on the brain. The surges in oestrogen and progesterone that occur during pregnancy, for instance, have been related to obsessive-compulsive disorder (OCD), an anxiety-related condition.

If you're experiencing anxiety, what can you do? If it's affecting your ability to live life or fulfil your potential, you can seek medical treatment. Cognitive behavioural therapy (CBT) is the gold standard, and aims to address the beliefs that drive your anxiety. If someone is worried about blushing, for example, they might be asked to put blusher on their face and have conversations with people to build their confidence. Or for an irrational anxiety about the fragility of your physical condition, you might be asked to run up and down the stairs to show that even if you do intense physical exercise, you aren't going to have a heart attack.

Recently, digital-based home therapies have become increasingly popular, to make up for the shortfall in face-to-face CBT professionals. Several studies show how powerful home-delivery of CBT can be – in some cases, just as effective as traditional techniques.[6] If therapy isn't your cup of tea, or if you're struggling to identify the source of your anxiety, another line of attack is to use drugs, which can redress chemical imbalances in the brain. Several studies show that people with high anxiety tend to have lower levels of a brain chemical called GABA, which is thought to help the amygdala filter out unthreatening stimuli. Benzodiazepines, a common class of anti-anxiety drugs, work on this brain system, boosting GABA levels. They take effect immediately but can only be used in the short term because they are highly addictive and lose their effectiveness over time. Longer-term options include antidepressants, which increase serotonin, a chemical that regulates mood. All of these drugs have side effects that can limit their use.

If you suffer from anxiety there are a couple of simple things you can do to reduce your need to seek medical treatment. The first is to exercise, which has been shown to be helpful in reducing anxiety.[7] During exercise endorphins are released, lifting the mood and enabling you to concentrate on something other than your own thoughts.

The second is to eat better. One study, from the University of Oxford, has found that taking a fibre-rich supplement to encourage the growth of beneficial gut bacteria for three weeks made people pay more attention to positive words on a computer screen and less attention to negative ones. Upon waking each morning the volunteers also had lower levels of the stress hormone cortisol in their blood, meaning there was a small but significant effect on the underlying psychological mechanisms that contribute to anxiety.[8]

So when modern life seems packed with events outside your control, apparently designed to foster the anxiety and self-doubt that erodes your confidence, make sure you can recognise the symptoms and act on them.

TOP TIPS FOR NOT WORRYING

- Ask yourself why you're worried about something – and then why you're worried about that, and so on until you get down to the roots of what is unsettling you.
- Deal with uncertainty by sitting it out – most of the time you'll find that nothing bad happens.
- Eat three weekly portions of oily fish, and plenty of seeds, kiwi fruit and leafy greens: foods containing omega-3 can help your body reduce excessive inflammation and de-stress.
- Work on nurturing your social ties, relationships and bonds with family and friends – it's thought to be the single most important factor in boosting your resilience to stress.
- For anxiety that won't go away try cognitive behavioural therapy (CBT).

2

HOW TO BE HAPPY

ONE OF MY favourite tricks I've learned is this: put a pencil between your teeth without it touching your lips. Hold it there for ten seconds and you should feel slightly happier.

Repeat this experiment in front of a mirror and you'll see why. Holding a pencil in this way makes you smile, which subsequently makes you feel happy. It's an example of how 'embodied cognition' can influence our emotions. Our brain infers how we are feeling in part from the movement of our muscles, so if we're smiling it's more likely to generate the appropriate associated emotion. We'll encounter embodied cognition again in the next chapter when we look at boosting confidence through body language.

The pencil trick is one I use all the time to boost my mood. It is one of many ways in which an understanding of how and why we feel happy can help us achieve more of it. And after the times we've been having recently, there's never been a greater need for a dose of cognitive sunshine.

HOW TO UNDERSTAND HAPPINESS

Happiness is not the easiest of subjects to study. For a start, scientists have never been able to agree on how best to measure it. Some researchers try to find answers by focusing on how happy someone is in the moment, and devise experiments to measure what's going on in the body and brain exactly when you feel a specific positive feeling. Others measure how satisfied people are with their lives in general, and focus more on the broader picture of what makes one population happier than another. Both approaches offer interesting results.

When I read research about happiness I'm often tempted to run away to one of the Scandinavian countries, which consistently come out top of the World Happiness rankings. What's their secret? In part, it probably comes down to a high GDP per capita, but then there are other similarly rich countries that fare less well, so that's not the whole story. Other key factors seem to be a fantastic state support system and strong community spirit. Job security and a good education also buffer against potential unhappiness. Even their tough winters and long nights might contribute, by bringing people together to help each other out. Or maybe it is all down to their cosy, 'hygge' lifestyles.

If such societal structures appeal, unfortunately it's challenging to implement them on our own. Aside from buying some candles and a soft throw, Scandinavia's happiness is difficult to emulate. So where else can we look in our pursuit of happiness?

One thing is certain: earning more money isn't the answer. I used to hate it when people told me this. Give me a million pounds and I promise you I will be happier, was my response. But most statistics contradict this instinct. Surveys consistently show that – crucially, once life's basics are paid for – family income accounts for only a small variation in happiness across the population. The power of money to bring happiness is limited.

Partly this comes down to something called the 'hedonic treadmill', neatly summarised by Aldous Huxley: 'Habit converts luxurious enjoyments into dull and daily necessities'. Basically, we get a hit of happiness from new stuff, but after it becomes a normal part of life that pleasure wears off and our happiness level returns to where it started. This makes sense: although the tendency to adapt to our 'new normal' constrains our enjoyment of pleasurable things, it also protects us from the impact of sad or harmful environments – we adapt to these too.

So if money isn't the answer, what is? Despite the disagreement about how to research happiness, many studies seem to have zeroed in on four aspects of life that are particularly influential. One is building your stress resilience, something we've already looked at. Another is increasing your mindfulness and teaching yourself to frame your thoughts positively, which I'm going to turn to in the two following sections. Then there is your environment, and finally your diet, which I'll come to after that.

HOW TO BE MINDFUL
AND LIVE IN THE MOMENT

When Daniel Gilbert, professor of psychology at Harvard University, discovered that people are much happier with irreversible than reversible decisions, it changed his life.

It might seem counter-intuitive, given that most of us would instinctively prefer to have options in life, like sending back an item of clothing we've just bought or choosing between four universities rather than having just one on offer. But actually lots of studies show we are more contented when decisions are set in stone.

In one of Gilbert's experiments, photography students believed that having the opportunity to change their minds about which prints to keep would not influence how much they liked their final selection. However, when tested, those who were able to change their minds liked their prints less than those who did not. It seems that we can rationalise a decision that's done and dusted, but not one that still has possibilities.[1]

On discovering this, someone mentioned to Gilbert that this was the essential difference between living with or marrying your partner. If your wife or husband does something mildly annoying, they said, you shrug your shoulders, but when your girlfriend or boyfriend does it you wonder if you should change partners. Gilbert immediately went home and proposed.

I'm not suggesting you do anything quite so dramatic to boost your happiness – but understanding some of the ways you can tame your unruly mind could help. Let's start with asking yourself a question: When are you happiest? When you're eating a delicious meal? Relaxing on a beach with your family? Having sex? One of Gilbert's colleagues, Matt Killingsworth, was interested in finding out the answer, so he developed an app called 'trackyourhappiness', which sent random alerts to thousands of people across more than eighty countries, asking them what they were doing and how they were feeling at that very moment.

One of the biggest surprises is that people seem to be happiest not when their mind is wandering, perhaps thinking lovely thoughts about the future, but when they are thinking about what they are doing at that moment, even when it's a boring chore like commuting or the washing-up. It may be one reason why social interaction makes people happy: you tend not to let your mind wander during these gatherings.

This analysis probably doesn't come as a surprise to anyone who practises mindful meditation. The art of focusing your mind on the here and now has been linked with increased happiness, alongside a host of other benefits. We've already encountered it in passing a couple of times in the previous chapter, but now's the time to look at the claims made of it in detail.

And if you're reading that last paragraph with a healthy dose of scepticism, excellent. You're right to be cautious. Mindfulness has gone from being a fringe topic to a replacement for psychotherapy, a tool for corporate well-being and even been touted as the key to creating more resilient soldiers on the battlefield. Of the hundreds of behavioural and neuro-imaging studies carried out on meditators over two decades, many are inconclusive. There is, unfortunately, a lot of misinformation and poor methodology associated with many studies.

There's even a small but concerning number of experiments that suggest that for some people mindful meditation can cause negative side effects like panic, disorientation and mania.

Thankfully, many scientists are now attempting to cut through this tangle of confusing claims and promises. One such neuro-scientist, Richard Davidson, and his colleagues have spent years collaborating with the Dalai Lama and other Tibetan monks. And while we still have a lot to learn, it seems there is much to be positive about. For example, there is good evidence that regular sessions of mindful meditation have a calming effect on the amygdala, the brain's emotion processor that was mentioned in the previous chapter. This reduces impulsive reactions to stressful or negative thoughts and experiences.

Studies also show it can help mute our emotional response to physical pain, and lessen anxiety and the kind of unfocused mind-wandering that is unhelpful to our happiness. The benefits are apparent even for beginners, and they increase with practice.[2] If you feel like giving mindful meditation a try, there are plenty of classes you can take or podcasts and audiobooks that will help. But you can start right now by closing your eyes and observing the present moment. The aim is not to empty your mind, utterly undistracted. It's about being aware of what's happening to you in the here and now. Think about your senses, what you can smell or hear. Some people like to do a body scan, thinking about how each part of their body feels. Acknowledge what emotions are present. If your mind wanders, thinking about stress at work, or some kind of life admin, make a mental note of it and let it roll on by. Draw your mind back to the present moment again and again, perhaps by concentrating on the breath going in and out of your body.

This isn't the only kind of meditation that seems to improve our well-being. Compassionate meditation, in which you consider how you feel about a loved one and think about extending this feeling to others, increases altruistic behaviour and makes us happier as a result. Studies show that when adults practise compassionate meditation for seven weeks, they not only feel a greater degree of positive affection afterwards, but also report greater joy, hope, gratitude, pride, interest and awe – and experience better relationships with others.[3]

Not a bad pay-off for a few moments of quiet contemplation.

HOW TO FRAME YOUR THOUGHTS POSITIVELY

If meditation isn't for you, there are other shortcuts to a happier mind. One is simply to think more about why you are thinking in a particular way. A few techniques from cognitive behavioural therapy can help with this. As we saw earlier, this therapy can be used to treat anxiety disorders, but it can also help to improve general happiness. You can even try to teach yourself the under-lying rules and put them into practice.

To do so, first you need to start paying attention to your internal thought processes. When you catch a negative thought, ask yourself, 'Is there really reason to think like this?' or 'Why am I feeling this way?' For instance, if you're worried about a future social meeting, thinking that you're useless at small talk, ask yourself what drives this negative thought and try to find a more realistic way to think about the event. Often just attending to your thoughts, rather than allowing them to drift in and out of your mind without conscious consideration, can help.

But to really get a grip on your cognitive processes, you need to restructure your beliefs, before challenging them in real life. You might, for instance, reframe your initial thought that you're bad at small talk in a more positive way – acknowledging that speaking with strangers makes you nervous, but that even a short conversation might benefit your career. Challenge yourself to approach just one person at first and then evaluate the outcome. It's sure to be more positive than you initially feared.

There's one more thing to try. Get out a pen and paper and write down one positive thing that has happened today. It might seem like some new-age mumbo-jumbo, but this simple act is backed by plenty of studies that show how writing about good experiences increases people's life satisfaction, with the benefits lingering for at least two weeks after the task.[4]

It doesn't have to be a long diary entry. During the coronavirus pandemic, this was the first tool I turned to, using Instagram to write about one positive thing that I experienced each day with the hashtag #littlebitoflovely. It not only helped me feel happier while I was posting, but I also felt it had an influence on my general well-being throughout the day, making me more mindful of the good aspects of my life as they were occurring. Of course, my experience is purely anecdotal, but it is supported by studies that show that people who write about their emotions for just a few minutes a day show fewer physical and mental health complaints weeks down the line.

Finally, give some thought to why you're doing all this. Happiness isn't just a perk of consciousness. One theory as to why it evolved is that it improves our cognitive capacities while we are in a safe situation, allowing us to build resources around us for the future. The idea is that when we feel happy, it changes the way our brains process our senses, expanding the boundaries of experience, allowing us to take in more information, to see the bigger picture. That's in stark contrast to negative emotions, like fear or stress, which focus our attention so we can deal with short-term problems.

When this theory is tested, by scanning people's brains or tracking the movement of their eyes while they have happy experiences, we see that a positive mood quite literally expands the scope of our visual attention, helping the brain gather more information from the environment. This has benefits beyond the initial burst of feel-good hormones that happiness gives us: feeling happy also improves your creativity and ability to solve problems.

For instance, in one experiment, volunteers were shown a video of comedy bloopers to lighten their mood, before being presented with a practical problem involving a box of matches, a box of drawing pins and a candle. They were told to attach the candle to a noticeboard in such a way that the wax didn't drip on the floor.* The experimenters found that people who had viewed the comedy clips were more likely to solve the

---

* The solution is to empty the box of drawing pins, use the drawing pins to attach the box to the noticeboard, place the candle in the box and light it with a match.

problem than people who saw a mathematics documentary intended to put them in a more neutral mood. Other studies have shown that a good mood improves verbal reasoning skills and social skills – making people more gregarious, trusting and able to deal with criticism more constructively. It all adds up to the idea that cultivating a positive mindset in good times helps us learn mechanisms that enable us to feel better in the bad times.

HOW TO FIND YOUR HAPPY PLACE

When I moved from a lovely, but cramped, Victorian flat in Queen's Park in north-west London to a mid-1960s wooden-fronted house at the edge of a leafy estate in Forest Hill in the south-east of the city, I was filled with apprehension. Would I be happy here?

I needn't have worried. My new community was diverse and welcoming, and had good restaurants, a small piece of woodland and plenty of wide, open parks. Within days I had been offered edging for my front lawn, regular curries from my neighbour, and a plot in a nearby allotment. But while the curries have been delicious, it is the parks and allotment that have made the biggest difference to my life. And studies show that in general having access to outdoor space is one of the most important ways we can improve our happiness.

So important, in fact, that in the Shetland Islands, the north-ernmost part of the UK, doctors are handing out some unconventional prescriptions. Along with regular therapies, people with a range of physical and mental ailments are told to take in the sounds and smells of seabird colonies, build woodland dens or simply appreciate the shapes of clouds. A similar scheme in New Zealand found that, six to eight months after receiving a 'green prescription', two-thirds of patients were more active and felt healthier, and almost half had lost weight. Elsewhere, this type of eco-therapy is emerging as a promising treatment for mild to moderate depression.

Scientists have long recognised that people living in greener neighbourhoods tend to have better well-being and lower mortality due to ill-health, even when different socio-economic circumstances are taken into account.[5] Last year, a study found that spending just two hours a week in green spaces boosts physical and mental well-being by about the same amount as getting enough exercise (that is in itself another way to boost your happiness, but you can find out more about that in Chapter 6). Research also suggests that the greater the biodiversity in green spaces, the larger the benefit.

Various explanations have been proposed. Firstly, being outside boosts your exposure to bright light. You may not realise it, but there's a good chance you're light-deprived. Standard office lighting tends to be around 200 lux. Even on an overcast winter's day, it's around 10 times brighter outside and on a sunny day it is up to 500 times brighter. Yet the average Westerner spends around 90 per cent of their time indoors.

Bright light stimulates your brain, boosting alertness and reaction times. It strengthens our circadian rhythms, the twenty-four-hour fluctuations in our biochemistry and behaviour that enable us to perform optimally, protecting us from depression and dementia. Connecting with nature also seems to provide additional physical advantages over and above the benefits of exercise. Merely being exposed to nature slows our heart rate, and boosts our immune response, and is probably related to a reduced stress response.[6]

But what if you're stuck in an urban jungle? There is another way to increase your dose. It turns out that you don't actually have to go into nature to reap some of the benefits. Experiments have shown that photographs, videos and audio recordings – 'surrogate nature' – have a similar, though less powerful, effect. Good results have also been reported with virtual reality. So if you don't have access to the natural world, look at pictures of it, listen to recordings of birdsong and other natural soundscapes, or maybe watch a David Attenborough documentary.

There is of course more to your living arrangements than the trees and forests. What greets you when you step through your front door? A clutter-free kitchen or a toy-filled monstrosity? A clean and tidy hallway, or messy shelves and dirty carpets? Recently, Japanese neatness consultant Marie Kondo hit the headlines with her 'life-changing magic of tidying up'. She promises that stress and inefficiency disappear with stray socks and the morning's dishes.

She may be on to something. For starters, a messy environment tends to make it harder to work. In lab experiments where volunteers are asked to do a simple task, they make more errors when the surrounding desk is messy than when it's clear. But it seems a clean environment can also have a significant impact on your mood. In 2009, researchers from the University of California asked men and women from different families to give a tour of their house. The team then analysed the frequency of words they used to describe clutter, and those that were more restful. Combining these, they created a scale to measure how stressful or restorative each person's home was. They then measured the participants' levels of the stress hormone cortisol, and their general mood over the following week.

People who had more unfinished, messy homes had flatter

cortisol levels, a profile known to be associated with adverse health outcomes, and had increased depressed mood over the course of the day, compared with people who had more restorative homes. This was particularly evident for the women in the household.[7] Perhaps it's no surprise that coming home from work and noticing piles of clutter and lists of to-do projects is associated with a depressed mood, but it's certainly something that is in our power to change. Clutter can also make it difficult to sleep. As we'll see in Chapter 7, that can severely affect your mood and health.

If you want to clear up your act you might be tempted by Marie Kondo's method: you take out every item in a room and dump it in a pile. You spend a moment inevitably overwhelmed by the disaster in front of you, before picking through each item, keeping only that which is useful or 'brings you joy'.

Whatever your preferred method of tidying, don't go too big, too fast. During the 2020 lockdown, I joined a small online movement called the #freshnestchallenge run by Hannah Bullivant, a writer and interior designer from London. The premise was simple: start small. Tidy one shelf. Clean a table. Spend fifteen minutes organising a drawer. Often you'll find you'll achieve more than you set out to. Bullivant may have developed this method of cleaning organically, but she accidentally hit upon the best way to get round one of your brain's natural stumbling blocks when it comes to keeping an orderly house.

That's because tidying up can give us a quick dose of the neurotransmitter dopamine, the pleasure chemical that boosts our mood. But this only happens if we set ourselves achievable goals. If the tasks we try to achieve are too big or too overwhelming, we will either not start them or fail to get any boost to our happiness when we do. Set yourself small goals, safe in the knowledge that the benefits go far beyond a tidy room.

HOW TO EAT YOUR WAY TO HAPPINESS

Nothing makes me happier than a huge bar of mint Aero chocolate. Throw in a glass of wine and I'm in heaven. But while researching this chapter I came across a study that made me rethink this statement. By getting people to keep track of their mood as they eat, it turns out that healthy snacks, fruit and veg or a lovely nutritious dinner can make us feel just as happy in the moment as our so-called comfort foods. This reveals something of a quandary. While we all instinctively feel that certain foods make us happy, when it comes to analysing how our diet affects our mental health, nutritional science gets very, very complicated.

One source of trouble comes from the way studies are carried out. They often focus on specific food groups, but people don't eat just one thing: they eat a mixed diet of components that interact in weird and wonderful ways. Add to that the complexity of the other lifestyle choices that influence our mental and physical health, and you've got a whole load of confounding factors that are difficult to control. A thousand people who eat pineapple and watermelon might have healthier hearts than those who don't – but tropical fruit is expensive, so it's just as likely that those who eat it are richer, have fancy gym member-ships, go on relaxing holidays and eat a whole lot of other healthier food. Was it really the pineapple and watermelon that protected them from heart disease?

Even when you try to compare groups with similar lifestyles, there are problems. People don't tend to keep accurate food diaries, accidentally or purposefully missing out certain items. Or they don't comply with the study rules, slipping in a slice of cake when they're meant to be avoiding sugar, for instance. Trials that strictly control people's diets are expensive to run and are rarely carried out. This all goes to say that if you're looking for a meal plan to make yourself happier, or healthier, you need to tread carefully. There's no quick fix, but the broad advice is that eating less fat, salt and sugar and more whole grains, fruit and vegetables will probably do you some good.

You might recognise this particular set of foods as the 'Mediterranean diet'. This approach to nutrition is holding up quite well compared with other passing fads. Besides being good for your physical health, plenty of studies show it might also increase your happiness, lower your risk of depression and boost your life satisfaction. Of course, as we've just seen, this might be in part down to people who follow this diet having other aspects of their life that make them happy. But growing evidence suggests this is not the whole story.

For a start, specific nutrients associated with the Mediterranean diet have many plausible effects on the brain. For instance, fruit and vegetables contain high levels of 'good' carbohydrates that are associated with a high concentration of serotonin, a brain chemical that increases happiness, positive mood and motivation. To produce serotonin we also need a healthy supply of trypto-phan, an essential amino acid that is found in dairy products, dried fruit and fish, as well as omega-3 fatty acids, magnesium and seeds, all of which are found in the Mediterranean diet. What's more, many nutrients found in the diet, such as B-group vitamins, play an important part in helping our cells generate energy, as well as being a source of vital antioxidants, which reduce cell damage.

But when you're thinking about dietary modifications to boost your happiness, you might also want to consider the small but mighty bugs that reside in your gut. From the day we are born, bacteria, viruses, protozoa and fungi colonise our guts, establishing a community known as our microbiome. Our microbiome helps shape our immune system, and heavily influences our physical health, mood and behaviour.

How does it do that? Some bacteria, such as *Lactobacillus rhamnosus*, which is used in dairy products, have potent anti-anxiety effects in animals. *L. rhamnosus* works by changing the expression of GABA receptors, which appears to have a calming effect on some areas of the brain. It's not completely clear how it communicates with the brain to enact these changes, but the effect is probably mediated by the vagus nerve, which connects the brain and gut. When this nerve is severed, no effect on anxiety or on GABA receptors is seen following treatment with *L. rhamnosus*.[8]

A similar thing might be occurring in humans. We know, for instance, that if you give people just the right combination of bacteria for six weeks, you can see changes in parts of the brain associated with mood, as well as improvements in the symptoms of depression[9] compared with people who take a placebo. Another study showed that a combination of *Lactobacillus helveticus* and *Bifidobacterium longum* reduced anxiety and depressive symptoms in healthy volunteers. Certain strains of bacteria can also reduce symptoms of irritable bowel syndrome, through a reduction in levels of the stress hormone cortisol.

Given the evidence, you might be tempted to guzzle a load of probiotics – bacteria-laden drinks and yogurt that promise to boost your 'good' bacteria. But here we need to sound a note of caution. Despite marketing claims to the contrary, there is

THIS BOOK COULD FIX YOUR LIFE

no real evidence that these kinds of products support any brain-boosting effects. Some bacteria fail to survive storage in the health-food store, or are eliminated by acidity in the stomach. Even if they do survive, they may not have the health benefits you seek. (That said, it is worth asking your doctor about what probiotic supplements might help if you're taking antibiotics, or suffer from constant sniffles. Some studies support the idea that they can reduce bouts of antibiotic-associated diarrhoea and ward off coughs and colds.)

Dietician Megan Rossi, who has spent years investigating our relationship with the bacteria in our guts, recommends a much simpler route to dealing with them. Different bacteria feed on different foods. Rather than concentrate on trying to make certain bacteria thrive for certain mood-boosting effects, we should instead be concentrating on increasing the diversity of the bacteria present in our guts. We can do this by having a wide range of plant-based foods in our diet.

We know, for instance, that people who have at least thirty plant-based elements in their diet in a week have a more diverse range of gut bacteria, which is associated, among other things, with better mental health and well-being. Plant-based foods are things like whole grains, nuts, seeds, legumes and fruit.

Rossi says there are easy ways to get your thirty elements. One simple idea is to buy a packet of mixed seeds and put a teaspoon on your breakfast: that's four elements right there. Or get a packet of mixed leaf salad, rather than a single lettuce, because each different type of lettuce has different plant chemicals that feed different bacteria.

There's one more benefit to putting healthy foods into your diet. Just weeks after I started trying to achieve thirty plant-based foods each week, I noticed a definite increase in my well-being.

Given the short time frame, it was unlikely to be down to any real effect on my gut bacteria, but it might not have been completely in my mind. Evidence suggests that making small changes in your life can lead to an increase in positive thinking and the self-perception of developing a healthy lifestyle, which alone can lead to more happiness.

So as long as those changes are evidence-based, you're on to a winner either way.

TOP TIPS FOR HAPPINESS

- Find ten minutes a day to practise mindful meditation. Studies show that this can quickly improve well-being, decrease the risk of depression and help control anxiety.
- Put pen to paper – people who write about their emotions for just a few minutes a day show fewer physical and mental health complaints weeks down the line.
- Tidy up! Cleaning your living quarters is a sure-fire way to a calmer, happier mind. Take it slowly – small tasks can give you a decent dopamine boost and are easier to achieve than trying to overhaul your entire home.
- Get outside. Spending just two hours a week in green spaces boosts mental well-being by about the same amount as getting enough exercise.
- Give some thought to what you're eating – a Mediterranean diet that includes thirty plant-based products a week increases the diversity of mood-boosting gut bacteria, and may be your best route to better mental health.

3

HOW TO BE MORE CONFIDENT

I WAS GIVEN the address of an office in the middle of Soho and told to give my name at the front desk. A man called David picked me up from reception and showed me into a second room, which was dimly lit with a large table in the middle. On the table sat a single radio microphone and a pair of headphones. I was told to take my coat off, put the headphones on and make myself comfortable. Then I was left alone.

Suddenly, over the headphones came the end of some music and a DJ's voice. 'Welcome back, you're listening to Heavy Radio and today we have Helen Thomson with us, who's going to be talking to us about her new book. Hi, Helen, how are you?'

To say I was startled was an understatement. Back in 2018, I had just written my first book, *Unthinkable*, about the strange highways and byways of the human mind. I had always suffered with crushing anxiety about any form of public speaking, and I had come to this office, I thought, for some very necessary media training before hitting the promotional circuit. Was this a real interview already?

I hardly had time for a nervous 'Hello' before the DJ jumped straight in. 'So you've written a book about weird people who have crazy brains – is it just a modern-day freak show?'

Needless to say, that wasn't quite the way I would have said it. I stumbled and spluttered my way through my answer, defensive and incoherent. Three questions later, I realised I hadn't even mentioned the book's name. It was the longest ten minutes of my life.

After the interview ended, David came back into the room, smiling and introducing himself properly. To overcome my confidence problems, he said, it was best to throw me in at the deep end with the worst possible scenario: a completely unplanned live interview.

By the end of that afternoon, I not only repeated that radio interview with clear-talking gusto, but sat in front of a camera for five minutes speaking eloquently about my book with only a glimmer of the previous, nervous author of earlier in the day. The difference was that I'd learned some tools to cope with my lack of confidence. Not just about what to say, but how to position my body, how to smile and how to reframe my thoughts. I had regained enough control of my confidence to keep it from slipping from my grasp when I needed it most.

That said, my first real interview was live on national TV on the *BBC Breakfast* show. Alone in the lift on my way there, I threw up. (Luckily, as I owned a dog at the time, I happened to have a spare plastic bag in my coat pocket, which saved any embarrassing mess.) But even though I walked into my first interview holding a bag of vomit, half an hour later I sat on set surrounded by lights, cameras and directors and gave what was (though I say it myself) a perfectly wonderful interview. Ever since, my interviews and public-speaking events have got better and better. I've even started to enjoy them. In the funny kind of way the human mind sometimes works, that first, car-crash fake interview had served its purpose. I noticed with each subsequent interview after that one how much better I was doing and that too became a source of additional confidence.

Confidence isn't just about public speaking. In our ancestral past, confidence may have been a simple but essential ingredient in the competition for resources – territory, food or a mate. These days, the ability to face challenges without worry and with confidence can have an impact on everything from our ambition, morale and credibility to our exam results and job promotions.

As my story shows, as much as we might describe ourselves as 'confident' or 'not confident', these things are not innate, immovable traits, but can be acquired and developed over time. So let's see how.

HOW TO UNDERSTAND CONFIDENCE

Scientific approaches to understanding what makes us confident vary. A neuroscientist, for example, might point you towards 'confidence neurons' – brain cells within our orbitofrontal cortex, an area that sits just above the eyes, which are particularly active when we make decisions.[1] There are actually many areas of the brain that become active when our confidence is tested – regions involved in tracking our expected performance on a task, or the pleasurable feelings we experience from making a good decision. When these areas are damaged, such as in Alzheimer's or Parkinson's disease, or after a stroke, people can become more risk-averse or even experience a dysfunctional sense of overconfidence.[2]

However, the most useful way to look at confidence is at the psychological level. For psychologists, confidence is a subjective feeling associated with our beliefs about the world: it is part of our personality. Psychologists tend to split personality into what they call the 'big five' traits: openness to experience, conscientiousness, extraversion, agreeableness and neuroticism (sometimes called emotional stability). The one that's particularly relevant here is extraversion, and the yin to its yang: introversion. Confidence is a primary component of an extravert personality.

Many of us want to encourage extraversion in ourselves and

our children. Such attitudes are understandable – a lack of confidence, or shyness, really can hold individuals back. Shy people tend to start dating later, have sex later, get married later, have children later and get promoted later, according to Bernardo Carducci, director of the Shyness Research Institute at Indiana University Southeast in New Albany. In extreme cases, shyness can even be pathological, resulting in panic attacks and social anxiety disorder, in which individuals shun public interactions.

To work out whether you're in the camp of the introverts or extraverts, answer some simple questions. Do you feel more comfortable on your own than being in a crowd? Are you very reflective? Do you prefer to retreat into your own mind for comfort or to rest, rather than talk to others? Do you prefer working by yourself to working in a group? Answer 'yes' to many of those questions and it's likely you're an introvert rather than an extravert. Most of us sit somewhere on a continuum between two extremes.

Again, a lot of neuroscience studies have attempted to distinguish how the brain works differently in people who are introverted or extraverted. For instance, people who are extraverted see a greater response in brain areas associated with attention and reward when they look at faces than introverts do. But it's pretty much impossible to make a causal link between these two things. It might be that extraverts get a greater feeling of reward from faces so seek out other people, but it's just as likely to be the other way around: that their enjoyment of being around people has led to the greater brain activity in these regions.

The most important psychological insight, however, is that your personality can change. This comes as a surprise, given the widespread belief that our personality is an integral and unchanging part of who we are. That idea came from psychologist William James, who in the late nineteenth century argued that our personality becomes 'set like plaster' by the age of thirty. Now we know that notion is incorrect. Personality in general is far more mutable than we thought.

While there's no doubt that some of our personality is down to our genes, our environment certainly influences it, even while we are still in the womb. There's evidence that mothers who are stressed during pregnancy are more likely to have an anxious child, for example. Experiences in childhood also shape our personalities. Young children become more extroverted and work harder when surrounded by other kids with these traits. Parental behaviour has an impact too. If parents encourage reactive infants to be sociable and bold, they grow up to be less shy and fearful.

None of that is perhaps any help to you now, but there's also no evidence that this moulding of our personality stops at thirty, or any other age. Indeed, there is plenty of evidence to contradict it. As we get older, we tend to become more agreeable, conscientious and emotionally stable. Another study looking at data from 4,000 people aged twenty to eighty found that personality is less stable in young adulthood, and also after age sixty. This makes sense if changes to the environment can influence your personality, since young and older adulthood are periods where people tend to experience maximum disruption to their environment.

I could go on: embarking on romantic relationships can reduce levels of neuroticism, the effects of unemployment makes people less conscientiousness and less agreeable, and people who are heavily invested in their jobs tend to show an increase in conscientiousness. Even moving to a new town or country might influence your personality – people living in New York tend to be highly neurotic, for instance, while Londoners score low on agreeableness (in that case, it may be that people with those traits are more drawn to those cities).

The extent to which environmental factors shape our dispositions over a lifetime is remarkable. When researchers compared results from personality tests taken by people when they were aged fourteen and again at seventy-seven, there was no evidence for stability in individual personality characteristics whatsoever.

And not only are our personalities not set in stone but we can also actively shape them. If this is your wish, you're not alone: almost 90 per cent of people would like to see at least a little movement in their ratings on each of the big five traits. If it is more confidence you're looking for, read on to see what works, and what doesn't, as we learn to harness the power of positive thinking. First, you might want to select an appropriate soundtrack.

HOW TO CREATE A
CONFIDENT ENVIRONMENT

Kanye West's 'Power' is what works for champion boxer Amir Khan. Tennis star Andy Murray prefers Ed Sheeran. For me, it's always been Bonnie Tyler's 'Holding Out for a Hero'. I'm talking about the music that pumps you up, that gives you a boost in confidence, whether you're facing a punch to the head, a Wimbledon final or a sold-out audience at the Hay Festival of Literature & Arts (here's hoping).

Scientists have known for a long time that music, and particularly loud music, can boost several aspects of your behaviour, including your confidence. The reason is no mystery: it's largely down to arousal. In this context, arousal refers to an enhanced state in which the body and brain are more alert, and emotions are intensified. Music can also distract our attention away from things that are worrying us.

The experience of power is vital to our confidence. When we feel more powerful, it increases what's known as our executive functions – things such as the ability to update mental information, to think abstractly, or see the bigger picture. In short, the things that make us more confident. Sometimes confidence is triggered by certain music because the two things are already associated in the brain. Sports music, for instance, may trigger the experience of power, because it is usually paired with competitive events associated with rewards and winning.

I first came across 'Holding Out for a Hero' when I watched and re-watched, again and again, the film *Short Circuit 2* in the late 1980s.[3] It is firmly paired in my brain with success and motivation. I still can't hear the song without getting a jolt of adrenaline. I used it to get through the hardest miles of the London Marathon in the early 2000s and I've never looked back. It gives me a boost of confidence every time I hear it.

Choosing the right kind of music has been shown to alter mood, enhance work output and make reaction times quicker. It can improve performance in many sporting and academic domains.[4] Be warned though, it also increases risk-taking.

You might already have a power song in mind. If not, what's the ultimate booster? When researchers in Hong Kong asked undergraduates to rate how powerful, dominant and determined each of a bunch of songs made them feel, they decided that 'We Will Rock You' by Queen, 'Get Ready for This' by 2 Unlimited and 'In da Club' by 50 Cent as the most empowering.

If music isn't your thing, there are other proven ways to tap into your inner empowerment and give yourself a confidence boost. One is to reach for a pen and paper. Psychologists in Germany hoodwinked students into writing about a time when they felt powerful, or a time they had lacked confidence, just before giving them a mock business school interview. Those who wrote about feeling powerful were much more likely to be offered admission by the panel than those who had written nothing. The worst performers were those who wrote about a time when they felt powerless.[5]

Then there's the curious effect of clothing. When researchers gave people the same white coat to wear, telling one group it was a doctor's coat, and the other a painter's smock, those who were dressed as doctors concentrated better on a subsequent task. It's a phenomenon the researchers dub 'enclothed cognition' and has been shown to enhance or diminish several kinds of emotions and behaviours, including confidence.

There are a few theories as to why positive thinking might be so effective when it comes to performance. Some of the most robust evidence comes from brain scans. For instance, when twenty-four women had their brains scanned during a maths test, those who had been previously made to feel empowered by exercises similar to the ones mentioned above not only scored better on the test, but experienced less activity in areas associated with 'cognitive interference'. Cognitive interference refers to thoughts and actions that interrupt the thing you're working on. These unwanted or undesirable thoughts can occur at any time and be fleeting, but suppressing them takes some cognitive effort, perhaps explaining why the women who experienced less of them were able to perform better on the test.

HOW TO BOOST YOUR BODY LANGUAGE

Pity poor old Sajid Javid. A lustrous career working his way through the ranks of UK politics and what he may be best remembered for is his power pose.

In 2018, the then Home Secretary was photographed with his legs further apart than you'd comfortably stand, arms gripped by his sides. The power pose. It later transpired that Javid was in fact mid-stride, turning to face the cameras, but he wasn't the first politician who had seemingly struck the pose. It sparked a national debate as to whether your body language can make you look – and feel – more confident.

The power pose was made popular by social psychologist Amy Cuddy. She encouraged us to copy what animals do to give the impression of power and dominance – make ourselves big, stretch out, take up space. Congenitally blind people do this just as regularly as sighted people at the end of a physical race, so it seems like something we've evolved to do. But would it work the other way round? Can we feel more confident merely by positioning our bodies in a certain way?

Cuddy certainly made a good case for it. She published a paper showing that people who were directed to sit or stand in certain power poses, with legs astride or feet up on a desk, reported stronger feelings of power than when they sat hunched over, legs crossed. The positions also changed their behaviour: power-posing made them more likely to gamble in a subsequent game. Even more compelling were Cuddy's physiological tests. Power-posing participants experienced about a 20 per cent increase in testosterone in their saliva, compared with a 10 per cent decrease compared with baseline for those who adopted the less powerful pose. Just two minutes of posing appeared to lead to hormonal changes that changed their brain to feel more confident.[6]

Later experiments would show that it worked outside the lab too. Cuddy got people to make strong or weak poses before a stressful interview. When other people watched the interviews afterwards, not knowing what pose the interviewees had adopted beforehand, they were more likely to hire those who had done the power poses prior to the interview. The interviewees' bodies had changed their minds, and their minds had changed their behaviour. They were putting the best version of themselves out there.

Cuddy's 2012 TED talk on this subject has been viewed 56 million times. No wonder, with such a simple premise. Unfortunately, over the next few years, things went rapidly downhill – not just for Cuddy's research but for psychology in general. A whole movement overtook the field, and a new statistical sophistication in methods of analysis raised the possibility that a huge amount of psychological research was unreliable. New standards of evidence were needed, more replication was essential and lots of previously accepted assumptions were now found lacking. Cuddy's research took a particularly bad hit and was heavily criticised by peers and the media.

One of the big problems with her research was that it didn't pass the p-curve test – a statistical tool that detects 'publication bias'.[7] In simple terms, it tests whether researchers may have caused errors in their data by cherry-picking certain data points most likely to produce a publishable result, or perhaps just got lucky with their data. The power pose didn't pass the p-curve test and was given the heave-ho.

In 2018, it made something of a comeback. Cuddy published an analysis of fifty-five studies that claims to demonstrate a real link between expansive postures and feelings of confidence and power.[8] It even passes the p-curve test, although no studies have been able to replicate the initial effect she found on hormones.

So where does that leave us? Even when factoring in the debate over power-posing per se, there are plenty of well-designed trials that suggest that how we use our bodies can affect our emotions, thinking and behaviour – and that we might be able to exploit these effects to feel more confident or worry less.

Take a minute to sit very still. Now, place your hands on the arms of the chair or desk in front of you, and try to focus your attention on counting your heartbeats. Can you feel the throbbing drum roll, the occasional murmur or nothing at all? How about your bladder? Is it empty or full? You might not realise it, but all of these sensations are helping you think.

Without these kinds of internal inputs, your mind would be unable to generate a sense of self or process your emotions properly. The physiological sensations from our bodies can influence everything from our tendency to conform to our ability to show willpower. Rather than the body being a passive vehicle driven by the mind, we now know that it is a partnership, with bodily experiences playing an active role in our mental life.

The most famous example of this is an experiment in the late 1990s in which Matthew Botvinick, then a doctoral student at Carnegie Mellon University in Pittsburgh, Pennsylvania, placed a fake arm where he could see it on a table, while hiding his real arm from view. Botvinick then asked an accomplice to stroke both rubber and real arms at equivalent places and in time with one another. In an attempt to reconcile the tactile and visual stimuli, he began to feel as if the stroking sensation was coming from the arm he could see. It was as if his brain had forgotten about the real arm and now felt it owned the fake one. He was suitably spooked by the sensation: he was so unsettled he threw the fake arm across the room.

Scientists call this embodied cognition. If you tried out the pencil trick at the beginning of this chapter, you may have experienced the phenomenon: the pencil made you smile, and therefore you felt happier. The same is true of other emotions: you might think you frown, say, because you are angry or annoyed, but in fact angry or annoyed feelings arise in large part from the physical sensation of frowning. Our ability to feel an emotion comes from the brain incorporating all of the environmental signals around us with the autonomic signals being generated by our body – our heartbeat, our hormones and what our muscles, limbs and facial features are doing. In one particularly elegant example, scientists showed that people whose frown muscles had been frozen with Botox took longer to read sad or angry sentences than they did before receiving the treatment. Without being able to express the emotions evoked by the sentence, they were less able to process the emotional language.[9]

The bottom line is that we know the body has a strong influence on the mind. So faking an air of confidence through your movements might well contribute to how you feel. Certainly, there's no evidence to suggest that power-posing and acting physically confident does you any harm. The best advice may be simply to consider how subtle influences around you shape how you feel about yourself. So stand tall, feet apart – just don't overdo it.

HOW TO BEAT THE GENDER CONFIDENCE GAP

There's an issue I've touched upon a couple of times already in this chapter, or at least seems implicit in a lot of the discussions surrounding something like power-posing. Now it's time to tackle it directly: the confidence gap, real or imagined, between women and men.

The wide gap in pay and opportunities between men and women in the workplace is often put down to an underlying confidence gap. The theory is that women are less confident in their abilities and are therefore not only less likely to come across as confident when in the limelight, but also less likely to put themselves forward for leadership roles, pay rises and promotions.

To some extent, there is evidence to support this. In a landmark study from 2003, David Dunning at Cornell University and Joyce Ehrlinger at Washington State University gave students a quiz on scientific reasoning. Before the test was handed out, the students were asked to rate their science skills. Women rated themselves consistently more negatively than men – they gave themselves 6.5 out of 10 on average compared with their male peers, who rated themselves 7.6. After the test, women also thought they'd done worse – they guessed on average that they'd got 5.8 questions correct, whereas men rated their ability at around 7.1. When it came to the actual results, there was no significant difference between their scores – with women and men both getting on average between 7.5 and 8 answers correct.

This lack of confidence from the women spilled over into the real world: before the results were announced, the students were invited to participate in a science competition that included prizes. Around 71 per cent of the men accepted the opportunity, whereas only 49 per cent of the women did. Less confidence meant less proactive behaviour in grasping opportunities.[10]

Sometimes women seem to do poorly because they don't give themselves the opportunity in the first place – scoring worse on tests just because they don't try to answer questions. When answering is required they tend to match men's accuracy. Men and women also react differently when asked to think about their confidence in an answer. Women tend to do worse after being asked this question, whereas the same nudge makes men perform even better.

All this suggests that women are walking around with less confidence than men. But it's bunkum: there's an equal body of research showing that in real life women feel every bit as confident in their abilities as men. So why doesn't it express itself practically in environments such as the workplace?

Part of the issue seems to be that a woman's self-confidence only translates into rewards at work when they combine it with empathy, altruism and other pro-social traits. Blow your own horn without these extra characteristics as a woman, and you're seen negatively. Men, on the other hand, can toot their trumpet and not be chastised.

To diminish this confidence gap, it seems more is needed than just women working on their self-confidence. Larger changes in the workplace are necessary: more transparency over who is responsible for what parts of a successful project, for instance, as well as making room for people to shout about their successes, would help normalise confidence for all.

There's significant evidence that positive role models can help, too. In 2013, social psychologist Ioana Latu, then at the University of Neuchâtel in Switzerland, and her colleagues recruited 150 male and female students to give a speech in front of a virtual audience. They found that men spoke longer – a standard indicator of confidence – and were rated more highly by an independent panel.

That in itself is not surprising, but the real interest came when Latu and her colleagues tweaked things for some participants by putting a photo of either Hillary Clinton, Angela Merkel or Bill Clinton in the back of the virtual auditorium.

It made no difference to the men. Latu speculates that could be because people in powerful positions, as men have often traditionally been, may be less influenced by outside information. But it made a difference to the women: the female students talked significantly longer and were rated more highly for quality when exposed to the to the images of the powerful women.

It's not yet clear exactly why this phenomenon occurs. But it helps to argue the case for having more role models, not just for women but for other groups such as ethnic minorities who may find themselves marginalised or lacking in confidence in a given situation. And it may be particularly important to have these role models visible in some way or another at the time you need a boost.[11]

HOW TO EMBRACE
YOUR INNER INTROVERT

We've talked a lot about the positive aspects of confidence, about how to overcome anxiety and how to cultivate our inner extraverts. But is there such a thing as too much confidence? One fact should give us pause for thought: we're not the only creatures to lie somewhere on a spectrum of introversion to extraversion. A similar phenomenon has been observed in spiders, sheep, birds and even sea anemones. It's clear that in the natural world, fortune doesn't always favour the bold: evolution hasn't selected for extraversion as the norm.

This suggests that introversion might have survival value. Some researchers think so, finding increasing evidence that people who have less confidence, who are shy, sensitive and even anxious – who are reactive, risk-averse 'pausers' – have advantages over more go-getting, risk-taking opportunity-seizers. Perhaps unsurprisingly, introverts are less likely to be hospitalised for accidents or illness.[12] Perhaps more surprisingly, a 2013 study of soldiers in the Israel Defense Forces suggests that anxious types are less prone to post-traumatic stress disorder following traumatic experiences in war.[13]

There are more benefits to being introverted. While it's been shown that introverts take longer to process information than extraverts, it's likely that they are processing it more thoughtfully. Introverts certainly interrupt less (more on interruption in Chapter 11), and their on-average superior listening skills can also be boosted by increased observation skills, what scientists call a 'sensory processing sensitivity'. Around 20 per cent of people have this skill. It's firmly linked with introversion, but also with an increased sensitivity to everything from music, art and novel situations to pain, drugs and coffee. People with this characteristic can detect small subtleties in the environment more acutely than others, performing much better than extraverts on tasks that involve using this talent.

And in fact there are distinct disadvantages to extraversion. Many of us routinely overestimate our abilities – a topic we'll return to in more depth when we discuss cognitive biases and smart thinking in Chapter 10. Psychologists have suggested that this could be an advantageous strategy because it increases our ambition, persistence or resolve and therefore generates a self-fulfilling prophecy. But it can also lead to unrealistic expectations and bad decisions. In fact it has been a bit of a puzzle among evolutionary psychologists how it could remain a stable strategy, evolutionarily speaking, among a population of people, some of whom have accurate, unbiased beliefs. You might have thought that natural selection would have weeded out this potentially harmful trait.

One computer simulation, developed by Dominic Johnson, an evolutionary biologist at the University of Edinburgh, and his colleagues, might have the answer. His team ran thousands of simulations in which generations of individuals made decisions based on varying levels of confidence in their assessment of others. They found that being overconfident was the most successful strategy when you are uncertain about your competitor's strength, and the benefit from the thing you're trying to achieve is sufficiently large compared with the cost of involvement. For instance, if you're competing with a work colleague for a project, and it's not clear who's the better employee, a little bit of narcissism and 'faking it till you make it' might be a wise choice, as long as the potential gains from being awarded the project are worth the risk of failing. Otherwise, caution should be advised.

All this work points to the idea that introversion and extraversion are both viable strategies in humans, and the relative success of each just depends on context. Given none of us know what tomorrow is going to bring, we cannot say that self-confidence and boldness are inherently good. In fact, children with genes that predispose them towards introversion are more developmentally malleable than their confident, gung-ho peers. Their future personality is more easily influenced by environmental factors such as upbringing. This means that when things go bad in life, they are adversely affected, but when things go well, they thrive.

The trouble is, in a society that so often sees boldness and confidence as king, shy people have a tough time. There are, of course, signs that the benefits of being introverted are being recognised. *Quiet: The Power of Introverts in a World That Can't Stop Talking* created quite a buzz when it was published in 2012, with author Susan Cain claiming that 'shyness and introversion may be essential to the survival of our species'. For sure, there may be times when you want to take advantage of some of the tricks laid out in this chapter, copy some notes from the extraverts' handbook and gain a little confidence. Just don't fall into the trap of regarding it as a sure-fire strategy for success in itself.

TOP TIPS FOR MORE CONFIDENCE

- If you can't make it, fake it. Stand tall, pull your shoulders back and hold your head high. Power-posing may give the impression of confidence to others, as well as making you feel more confident yourself.
- Listen to 'We will Rock You' by Queen, 'Get Ready for This' by 2 Unlimited and 'In da Club' by 50 Cent before a big event. These were rated as the most empowering confidence-boosting songs.
- If you are female, surround yourself with female icons that you respect – it can boost your feelings of empowerment and improve the way you come across to others.
- Positive thinking works. Just before your big event, write about a time you felt powerful or had a lot of confidence. People who do this are more likely to be successful than those who write nothing.
- Remember, fools rush in where angels fear to tread: sometimes over-confidence can lead to bad decision-making, so embrace your inner introvert when a decision has a lot at stake.

4

HOW TO MAKE FRIENDS

WE SAT IN THE upstairs room of a pub. Five couples, all nervously smiling at each other, trying to make small talk. We had just two things in common. We lived near each other, and every one of us was weeks away from having our first baby.

I was at my first antenatal class. I knew these women, and their partners, could potentially be my friends for life. At the very least, they were to be the support I would need in those first few months of survival with a newborn. The pressure of making a good first impression, of saying the right thing, of not wanting to be sidelined by the group was immense.

Five million middle-of-the-night WhatsApp messages later (I'm exaggerating, mildly), and those four strangers are among the people I talk to the most, who I turn to in times of crisis and without whom I would have suffered far more, mentally and physically, through the first years of my daughter's life.

We all need friends. Don't take this statement lightly. Friends are not just necessary for keeping our social calendar full. They improve our mental health, keep us physically fit and even prevent us from dying an early death. Social isolation, on the other hand, leaves us feeling physical pain, stress and makes us susceptible to illness. In fact, if we don't have friends, our bodies react as if a crucial biological need is going unfulfilled.

Even so, making friends, especially later in life, is not always easy. Perhaps you have lots of friends, but would like a few more, or different ones. Or you're moving to a new city and starting from scratch. Maybe you're feeling incredibly time-poor and want to know how best to invest in the friends you have.

Breaking the ice, spending time together, navigating social etiquette – these things don't always come naturally to us. And it takes time to make friends. A study from Kansas on adults who had recently moved home and were looking to make new friends found that it takes fifty hours of time spent together for you and another person to become casual acquaintances. Ninety hours and you might think of them as a friend. At least two hundred hours spent together and they could be your closest buddy.

It may not be possible to speed up the process, but there are things that science can do to lend a helping hand. Scientists can't give you an instant BFF – a 'best friend forever', but they can explore the impact of friendship on your life, how best to maintain the friendships you already have and how to break the ice with new people.

And ultimately this can go some way to prevent one of the most neglected public-health issues of our time – loneliness. A 2018 study suggested that about nine million people around the UK are lonely. Getting out of a cycle of loneliness is not simple. Even people with plenty of friends can often still feel lonely. But there are ways to tackle the problem. For selfish reasons, it's more than worth the effort to understand all of this – not only will you become a better friend to others, but also it's a sure-fire way of boosting your own health, increasing your lifespan and guaranteeing a lifetime of happiness.

HOW TO UNDERSTAND FRIENDSHIP

It might seem a bit weird to relate friendship to breastfeeding, but that's where our love of friends might start. As a baby suckles at a breast, oxytocin is released from the mother's pituitary gland. This causes muscles in the breast to contract, allowing milk to flow, but it also reduces anxiety, blood pressure and heart rate. For mothers and babies, the related feeling produced by oxytocin encourages suckling and helps create a strong and loving bond. This occurs in all mammals, but in humans and other species that make friends the system has been co-opted and expanded.

Rather than reinvent the wheel, evolution has economised: oxytocin has become associated with relationships beyond the mother–child bond. You release it in response to many types of positive physical contact with another person, including hugs, light touches and massage. The resulting pleasant feeling is your reward for the interaction and encourages you to see that person again. A budding friendship is born.

It's not the only chemical driver of friendship. Our friends also trigger the release of endorphins, which create a feeling of well-being. When we spend time working on activities with friends we release more endorphins than doing that same activity alone. So good is this feeling that when we look at pictures of our friends in a brain scanner, brain regions associated with addiction light up with activity.[1] Friendship is driven by a similar system of reinforcement and reward that we see in people who smoke or take drugs.

Also triggering the release of endorphins are several additional behaviours that can be done in groups, allowing several individuals to bond as friends at the same time. Wonderfully, one of the ways we've done this is through laughter. Then there's singing, dancing and finally just language, all of which increase the opportunities to connect with lots of people at once.

Although we all have these same neurobiological processes, some people are better at making friends than others. Maybe they simply have the talent to come across as someone you want to spend time with, but maybe they're also more motivated to do so because it gives them a bigger chemical kick. We do know that friendlier people are more sociable, in part because their genes make them that way. When researchers compared the social networks of identical twins, who share all their genes, and paternal twins, who share around 50 per cent, they found that genetic factors accounted for 46 per cent of the differences in how popular among their peers individuals were.[2]

But even the social butterflies among us aren't friends with everyone. Of the many people we encounter in our daily life, how do we pick a select few as friends? The answer, at first, seems quite simple – we are friends with people who are similar to us. They're a similar age, a similar gender, or have similar jobs and moral outlook. But it turns out that this tendency to be drawn towards people who are like us also has to do with our genes. In fact, you are as genetically similar to your friends as you are to your fourth cousins.[3]

This offers an answer to one of the great mysteries of friendship. Cooperation is vital to our health and well-being, but why would we cooperate so readily with complete strangers? In evolutionary terms, you should cooperate with kin, rather than kindred spirits, because your genetic similarity to relatives allows you to reap indirect benefits. You succeed by proxy if they pass on more of the genes they share with you to future generations. If friends are more genetically similar than we would expect by chance, perhaps they're not so much strangers as 'optional relatives'.

How we recognise genetically similar people is another matter. Various studies have suggested that it may be down to similarities in facial features, voice, gestures, smell or all of the above.

Your personality is shaped by your genes, so perhaps choosing friends with a similar personality means they're more likely to have genes in common.

Whatever attracts us to someone, one thing is certain, befriending them will be rewarding. We don't need the films *Thelma and Louise* or *I Love You, Man* to highlight the importance of friendships in one's life. Neither do we really need to turn to the decades of empirical research to be in no doubt that friendship is associated with happiness and good health – a finding, incidentally, that is consistent across all gender, ethnic and cultural groups. But if we do, we find that people with weak social relationships are 50 per cent more likely to die in a given period than those with strong social ties. Social isolation, as we'll hear more about later, is as bad for you as drinking or smoking – by some estimates, it's equivalent to smoking fifteen cigarettes a day – and worse for your health than inactivity or obesity.

A rather less clear-cut question is how many friends we need, or indeed can have. Monkeys and apes have an upper limit of around fifty other animals in their social group. Our group-bonding behaviours mean we can do a little better. Robin Dunbar, an evolutionary psychologist and friendship expert at the University of Oxford, says that on average humans have five intimate relationships – he calls these 'shoulders to cry on'. We also have fifteen close friends and family, fifty good friends (the ones you would invite to a party) and a hundred and fifty people who might turn up to your funeral.

But the number of friends you have really doesn't matter – it's the quality of the friendships that counts. We should, for a start, beware of 'frenemies'. Interacting with these unreliable friends is stressful: your blood pressure is more likely to be elevated when you are with a frenemy than with someone you do not like at all.

In general, the quality of a relationship depends on how much time is invested in it. With only so much time available, if we try to invest in too many people, the quality of the friendship is poor. Our social groups are less stable, and more liable to break up. According to Dunbar, to maintain a friendship you need to be in contact with very close friends about every other day and your next five closest pals about once a week, whether face-to-face or electronically. Once a month is enough for the next fifteen. For the other fifty it's about every six months, and for the rest of your hundred and fifty or so connections, once a year. Less often than that, and friends will quickly fall through the layers of your social network. The exception is close friendships forged in your late teens and early twenties – you can often pick these relationships up exactly where you left off, even after decades apart.

There's no sure-fire way of creating these friendships. Most form organically as we attend school and groups as children and get jobs and socialise as adults. But if you're interested in making a few more, the best thing you can do is put yourself in a place with people with shared interests.

Studies of friendships suggest there are six important criteria we share with our friends: our language, profession, world view, sense of humour, local identity and education. Personality appears to be less important than cultural preferences: the bands you like, the books you enjoy, the jokes you find funny. In fact, heading to a gig or joining a music club, band or choir might be one of the easiest ways to make new friends, since the best predictor of how well you will get on with a stranger is whether you like the same music.

Which takes us nicely on to the next step – making the first move.

HOW TO MAKE SMALL TALK

'Conversation should be like juggling,' wrote Evelyn Waugh in *Brideshead Revisited*. 'Up go the balls and the plates, up and over, in and out, good solid objects that glitter in the footlights and fall with a bang if you miss them.'

Do you feel that you always manage those linguistic acrobatics with aplomb? I must admit I don't – I feel like I'm walking through life permanently with my foot in my mouth. I often leave a conversation worrying that I might have said the wrong thing, should have listened more, or could have come across better.

The good news is that we're probably more proficient conversationalists than we think. Studies show that we consistently underestimate how well we are perceived by others, a phenomenon called the 'liking gap'. But whether we're attempting to make friends, or merely being forced into a social situation in which we don't know many people, psychology can also offer us some tips on the art of conversation – from the etiquette of eye contact to the most tactful exit plan.

Let's start with that first impression. Actually your words matter less than the general backdrop of feelings that you and your potential future friend may be having. Aspects of that might be out of your control: even the weather can influence someone's opinion of you. But here's how not to worsen someone's mood. Distance is your first consideration. We've all met people who get a little too close for comfort, but you don't want to be too far away either. The optimum space while chatting, according to a study of nearly 9,000 people in forty-two countries, ranges from about 40 centimetres to three times that, depending on your culture and the nature of the relationship between you. In the UK, about 99 centimetres is preferred for a complete stranger – about the width of a kitchen table. If you happen to be making friends in Romania, however, be sure to keep any new associate at much more than arm's length, as people there generally prefer to leave around 140 centimetres between strangers.

We've already seen how we tend to prefer people who are similar to us, and one way of signalling this is by mimicking. Mirroring someone's posture and facial expression can help to nurture a shared bond and increase cooperation between people. Subtly imitating someone's voice, by matching their pitch or speed of speech can also work in your favour. Even repeating or paraphrasing someone's words back to them can create a sense of bonhomie. Just don't make it too obvious.

Next, look into their eyes. This helps us to read emotions more accurately and signals that we're listening intently. But again, be careful. While avoiding eye contact can make you seem untrustworthy, too much staring can be unbearably intense. If you really want to get into the nitty-gritty, 3.3 seconds of eye contact is about how long we can stand before feeling uncomfortable.

Once you're settled into the conversation, don't forget to ask questions. In first encounters particularly, our default behaviour seems to be to want to talk about ourselves. In fact, the number of questions you ask of someone during a conversation can reliably predict how much they like you afterwards. Studies of speed-dating events show that it also predicts how likely you are to agree to a second date.

The type of question matters, too. 'Switch' questions, which alter the topic of conversation, are less charming than follow-ups that build on the person's current topic. It signals a kind of emotional responsiveness and care for the other person.

You can tell whether it's going well by judging how easily the conversation flows – the seamless turn-taking that gives the impression of having clicked with someone. The shorter the gaps between turn-taking, the greater the rapport reported between individuals. Slight pauses break up this rhythm and make us feel uncomfortable, like we haven't been appreciated or understood. Perhaps this is why conversations via video conferencing apps, with intermittent delays, can be so excruciating.

One of the most important things to know about a good conversation involves interrupting. Nothing can appear as rude as the tendency to speak over other people. It not only affects your ability to build a relationship with the individual being interrupted, but actually reduces the overall collective intelligence and problem-solving ability of any wider group you may be in. Interruptions aren't just verbal. Checking your phone is also detrimental to a conversation, and even the presence of it in your hand or on a table can reduce empathy between you and your conversation partner. If you want a meaningful conversation, best just put it away.

Yet interruptions can serve a useful purpose if they are used to maintain rather than disrupt the conversational flow. You might interject to finish someone's thought, for instance, particularly if they are already trailing off their sentence, since that helps to avoid an awkward gap in the dialogue. One study of speed daters found that these interruptions were associated with a greater sense of connectedness and understanding between the speakers. So don't be afraid to step in occasionally if it helps maintain conversational momentum.

Finally, if you've tried your best, but you're stuck with less-than-agreeable company, with no common ground, you can use the flip side of these same insights to put a gentle end to the discussion. Signal your disapproval without open confrontation by means of a few short but slightly awkward pauses that speak for themselves. Alternatively, change your body language, creating more distance, or simply place your phone on the table to indicate your interest is waning.

Give it all a go and see what works for you. You never know, getting over your fear of schmoozing could lead you to your next best friend.

HOW TO MAINTAIN FRIENDSHIPS WITH TECHNOLOGY

US President Woodrow Wilson called friendship the only cement that will hold the world together. A century on, and you often hear the fear voiced that our fast-moving, high-tech and increasingly urbanised existence is causing that cement to crumble.

Those fears are not without cause. Since the 1980s, the average US citizen's number of closest friends fell from three to two, and the proportion of people with no confidants at all increased from 8 to 23 per cent. In the UK, a rise in the number of people living alone and the weakening of community ties have led to warnings of a crisis in friendship. Other studies have linked the internet and mobile phones with increased social isolation.

Social media certainly has the potential to become antisocial media, but there is more evidence that it is changing the traditional notion of friendship. We still have our core group of friends, the ones we hang out with most, online or offline. The most obvious difference, however, is in the number of people we have some kind of contact with. In 2014, a typical teenager had around 300 Facebook friends and 79 followers on Twitter (not all of those would count as social ties, as they may not be being followed in return). This is far more than the maximum number of 'meaningful friends' our brains have evolved to deal with.

These extra people are known as weak ties: high school or college friends, work colleagues past and present, casual acquaintances, occasionally complete strangers. Social networking allows us to maintain a relationship with these peripheral friends that would never have occurred otherwise. But technology does more than that. It can improve our most fragile relationships, with those people with whom you might not be able to engage any other way. The further you live from your friends the more you engage online. It may be the difference between a real relationship and the memory of one.

You might be bored with Facebook, or think these random comments – wishing someone you barely know a happy birthday, or liking a picture on Instagram – are slightly vacuous, but they are all ways in which you are signalling that you are paying attention to that person. The deep, emotional bonds that characterise our most important relationships are still mostly cultivated face to face, even if we nurture them online. But these actions are still a highly effective way of maintaining weak ties, and there are many good reasons to do so. Weak ties tend to be diverse and span different social groups, so can provide new ideas and perspectives, an incentive for innovation, job openings and the sense of being part of a wider community.

One of the most dramatic illustrations of how the reach of our social networks affects us day to day is the ease with which we soak up the moods and emotions of people we don't know well. This has always been the case in the real world – you see someone smile and you smile back. But in online networks, this contagion effect is amplified. An analysis of more than a billion status updates on Facebook revealed that people inadvertently transmit positive and negative moods via their written comments, even to friends and acquaintances living in different cities – their weak ties. Now more than ever, we feel what the world feels.

The landscape of friendship has certainly been transformed in the past decade, but whether this has been for the better is still hotly debated. Some studies indicate that interacting with people online is just as valuable psychologically as interacting in person, reducing anxiety and depression and increasing feelings of well-being. In one study of Facebook, the more people use the site to actively engage with their friends, the less lonely they are – although it is not clear whether Facebook use reduces loneliness or whether people who are already socially connected use it more.

However, there are risks to maintaining friendships online. The nuances of interactions can be lost, for one. What might have been a relatively meaningless comment to a friend over coffee, misconstrued and then clarified in a matter of seconds, can now be an enduring statement, seen and misinterpreted by many. Also, modern-day social networking fosters narcissism and individualism, and if we disclose our emotions to more people we leave ourselves open to being hurt by a larger number of people.

The lasting thread through all of this is thinking about quality, not quantity, of friends. A large social network provides you with plenty of opportunities to make contacts or gather information, but when it comes to feeling a sense of warmth, belongingness, support and trust – the key pillars of happiness – having a few close friends is what matters.

Even here, however, technology might help. Over the past few years, many apps have been launched to help people manage their relationships better. These tools promise to help us stay in touch in a more meaningful way, to be more thoughtful friends. If it sounds tempting, you're spoiled for choice. From the least to the most inexplicably named, you can now download Ntwrk, UpHabit, Plum Contacts, Dex, Garden, Levitate, Monaru, Clay and Hippo. Most work in a similar way: you import your contacts to the app, label and tag them, as friends, family or co-workers, for instance, set reminders for when to contact them and log the topics you last spoke about. In theory (although this has not yet been tested in any significant study) this means you will never forget the name of an acquaintance's kid or to ask your uncle how his knee surgery went, and you'll have stronger, better relationships as a result.

Alongside reminders about birthdays or other important events, the apps send notifications to offer help to people, remember to ask about an interview, or check in with a simple hello to those you might not have spoken to for a while. When a colleague of mine at *New Scientist* spoke to people who used them, she found many who swore by them. For instance, Timothy Luoma, a Presbyterian pastor in Plattsburgh, New York, uses UpHabit to remember the details of his parishioners' lives – in particular to keep track of death anniversaries. 'The overall goal is to make sure that I am taking good care of my entire congregation, and not just those people who I see regularly, or the squeaky wheels,' he said.

Personally, I'd like to think that a good friend wouldn't need

THIS BOOK COULD FIX YOUR LIFE

an app to remember that I'm due to give birth in a few weeks or that I'd got a new job. But it's easy to see how these apps can suit some niche needs, for people who struggle with memory loss, say, or for maintaining social connections at times that are particularly busy. And the risk of not maintaining those friend-ships might outweigh the risk of seeming shallow. So perhaps it's worth a try. You can always uninstall an app, but it's much harder to reinstall a lost friend.

HOW TO TACKLE LONELINESS

Imagine you're a zookeeper and it's your job to design an enclosure for humans. What feature would best ensure the health and well-being of the people in your care? After food and water, there is only one answer – other humans.

Loneliness can be one of the most toxic environmental condi-tions we can encounter. It changes the brain, taking hold of our thoughts and behaviours in ways that are likely to make us feel even more isolated. Its effects aren't just psychological, but physical. Left unchecked, loneliness can be as detrimental to longevity as smoking, alcohol and a lack of exercise. People who are lonely are at increased risk of just about every major chronic illness – heart attacks, neurodegenerative diseases, cancer. It increases the odds of an early death by 26 per cent. That's about the same as being chronically obese.

Loneliness is often assumed to be a problem of social isola-tion, one that predominantly affects the elderly or vulnerable. There is some truth to this: half of people who are of retirement age in the UK say that the television is their main source of company. But loneliness can affect anyone. You might have experienced it yourself during the recent coronavirus lockdown.

Actually, loneliness has less to do with being on our own, or having few friends, even if that is how it is often defined. It's not social isolation, it's feeling socially isolated. A lonely person will not feel less so simply by being surrounded by others. Similarly, a social butterfly won't feel lonely just because they spend some time alone.

One apparent reason for the detrimental effects of loneliness is that it lowers our willpower so we're more likely to indulge in self-defeating behaviours (for more on this, see Chapter 8). We may take more risks, make worse decisions surrounding food and exercise, and have an increased risk of anxiety, stress and depression, all of which have a knock-on effect on our physical health.

The damaging effects of loneliness may also relate to a lack of sleep. People who are lonely are more likely to feel tired in the day and sleep fitfully at night. Disrupted sleep, as we'll see in Chapter 7, is a known risk factor for many chronic illnesses and mental health problems.

Why, then, would we have evolved to feel such an emotion? The most likely interpretation is that loneliness helped us to maintain the cooperation with others that is vital to our survival and reproductive success, whether it's fighting a predator or finding food. For our ancient ancestors, a short pang of loneliness was part of a biological warning system, like hunger or pain, that calls on us to change our behaviour, seeking out safety in numbers.

So why are some people more prone to loneliness than others, and is there anything we can do to avoid it? Studies of twins suggest a genetic component predisposes some people to a greater need for strong social connections, but environmental factors clearly play a vital part. Young people today seem particularly vulnerable to loneliness, for instance, with some researchers suggesting that this can be attributed to people moving around more, so ending up with fewer close relationships and social support in local communities.

If you're feeling lonely, what can you do? First, you should be aware that loneliness can trigger a vicious cycle. This is in part down to the fact that people who are persistently lonely seem to have the activity of genes responsible for inflammation ramped right up. As a protective measure to our health this has benefits: inflammation triggers behaviours like vigilance and suspicion, and dampens down brain areas involved in motivation to see people. This is helpful if you're sick and need to isolate yourself and rest. But in the modern world, it can make us bad at reading social situations and prevent lonely people from seeking out the company they need. It can be a cycle that easily spirals out of control.

If there is one thing that can help, it is to work on the quality of our friendships. To prevent loneliness, we have to have a close group of friends we can rely on to be there for us. It's hard to put a statistic on such things, but studies suggest that around 40 per cent of your social effort should be focused on this close circle, meaning you need to see or interact with them on a very regular basis. Small changes like pruning random acquaintances from social media, setting notifications for updates from real friends and spending time with a core group can all act as a buffer against loneliness.

If loneliness has taken hold, there are things you can do to escape. An analysis of interventions to reduce loneliness found that one of the most successful was cognitive behavioural therapy, something we encountered in previous chapters when we discussed anxiety disorders and developing positive framings for your thoughts. The heightened sense of threat lonely people feel means they are more likely to pay attention to and remember negative details and events, and behave in ways that confirm their negative expectations, perpetuating the vicious spiral. Likewise, a separate study showed that finding a sense of purpose and meaning in life can help overcome the negative effects of loneliness.

It sounds counter-intuitive, but fixing loneliness might be about helping others, rather than helping yourself. If you think of lonely people as having a world view of threat and hostility, you need to attack this underlying psychology by becoming engaged in helping others, trying to make the world a better place. It sounds like hard work, but the pay off – for yourself, and for your friends – is huge.

HOW LONELY ARE YOU?

Answer the questions using a scale from 1 to 4, where 1 = never, 2 = rarely, 3 = sometimes and 4 = always, then calculate your total score.

1. **How often do you feel unhappy doing so many things alone?**
2. **How often do you feel you have no one to talk to?**
3. **How often do you feel you cannot tolerate being so alone?**
4. **How often do you feel as if no one understands you?**
5. **How often do you find yourself waiting for people to call or write?**
6. **How often do you feel completely alone?**
7. **How often do you feel unable to reach out and communicate with those around you?**
8. **How often do you feel starved of company?**

9. How often do you feel it is difficult for you to make friends?

10. How often do you feel shut out and excluded by others?

How you scored

20 is the average score on this survey

25 or higher reflects a high level of loneliness

30 or higher reflects very high levels of loneliness

Source: Daniel Russell, UCLA

TOP TIPS FOR MAKING FRIENDS

🔧 Take your time. It takes about ninety hours spent together to forge a friendship. And at least two hundred hours to become best pals.

🔧 Put yourself in a place with people who have shared interests. Heading to a gig or joining a band or choir might be one of the easiest ways to make new friends – the best predictor of how well you will get on with a stranger is whether you like the same music.

🔧 Become an expert at small talk. Subtly mimic the other person's tone and actions, don't interrupt unless to support their point, look them in the eye often and ask lots of questions.

🔧 Focus on quality rather than quantity. Spend at least 40 per cent of your social interactions with your closest circle to maintain the support you need and avoid the detrimental effects of loneliness.

🔧 If you feel lonely, consider cognitive behavioural therapy.

Loneliness can heighten your sense of threat, which makes you more likely to pay attention to and remember negative details and events, and behave in ways that confirm your negative expectations. CBT can help you to get out of this vicious spiral.

5

HOW TO
FIND LOVE

WELCOME TO A chapter written by a woman who's had her heart broken on several occasions, who once had to sneak a peek at a driving licence to remember a date's name, and who went on dates with twenty-four people she met online before meeting the man of her dreams. Even calling someone that makes my skin crawl slightly. I don't believe in love at first sight, I don't believe that everyone has one person for them, and I'm not sure I even believe that monogamy is the best choice for everyone. Not the classic CV of a love guru.

But as you've probably noticed by now, I do believe in science. It might not be the most traditional bedfellow of love, but over the past fifteen years I've been writing about the brain and behaviour I have come to the realisation that it has made me an accidental expert on how to find – and keep hold of – a lasting, loving relationship. If only it had dawned on me earlier.

It turns out that neuroscientists, behavioural psychologists, evolutionary biologists and even geneticists all have something to contribute to our love stories. Ever since Charles Darwin outlined his theory of sexual selection, scientists have been fascinated by the way animals compete with one another to find a partner. By understanding what happens in our brain as we fall in love, how that changes throughout our relationship, how our innate biases might help or hinder our attraction to someone, and how our language can make or break a marriage, we can all do better in the game of love.

Granted, tests of attraction in the lab are hardly a replica of how we feel in real life, and I dare you to show me a perfect relationship to put under the microscope. But after spending so much time wading through the best and worst experiments and ideas, I can round up the best advice on finding love, from the best colour to wear on a first date to the way to solve an argument after twenty years of marriage. And for when things go wrong, I can also reveal the four most corrosive behaviours that predict divorce and a few weird and wonderful ways of helping you mend a broken heart.

As Shakespeare wrote, the course of true love never did run smooth. But you can definitely sidestep a few of the bumps.

HOW TO UNDERSTAND LOVE

What is love? Excellent question. Shakespeare (him again) said it was 'an ever-fixed mark/ That looks on tempests and is never shaken'. For neuroscientists, it's a little less poetic: they talk of love as a neurobiological phenomenon that can be separated into three behaviours: lust, attraction and attachment. All of these evolved to increase our reproductive and parental success. While Shakespeare's definition may appeal more, I think it's worth taking a quick look at what happens in our brain as we fall in love, because it can help explain a lot of the weird feelings we get at different stages of a relationship.

Lust, attraction and attachment are all grounded in a suite of overlapping chemical systems in the brain. Let's start with lust, the emotion that has led many of us to obsess over the tiniest detail of a person's smile, the way they laugh, or the number of kisses they send in a text. It's this kind of tunnel vision that led Donatella Marazziti, a professor of psychiatry at the University of Pisa in Italy, to compare lust to some of the symptoms of

OCD. When she compared the brains of twenty people in the first throes of love with those with the disorder, she found many similarities. Both groups had unusually low levels of a protein that transports serotonin around the brain. Serotonin does an incredibly important job in regulating our mood, happiness, sleep, anxiety, appetite and sexual desire. No wonder we act a little strangely when we're lusting over someone. When Marazziti retested the lovers a year later their serotonin levels had increased; they also no longer felt an obsessive need to focus on their partners.

This kind of intense attraction, and the accompanying euphoria, craving, withdrawal and relapse that we often experience at the beginning of a relationship also resembles another condition – addiction. Brain scans of people in love show a lot of drug-addiction-like activity in the brain's reward centres. It's not surprising, really, given that the early stages of romantic love is thought to be an uncontrollable drive that evolved millions of years ago as a survival mechanism to encourage bonding. It's an extreme version of the reward mechanisms we have already seen kick in when we make friends.

All this craziness seems to settle down over time. When you peek inside the brains of couples who have been married for more than two decades they reveal a different story. Thankfully, this group still shows lots of activity in areas of the brain associated with motivation, 'wanting' and pleasure, suggesting that the rewards associated with long-term partnership can be sustained in a manner that is similar to young love.

But the group also showed unique activity in the striatum (a brain area we'll encounter again when we examine how we form habits and addictions in Chapter 8), which is important in goal-directed behaviours. While it's always difficult for neuroscientists to understand how specific brain activity relates to behaviour, the accepted interpretation of this finding is that it reflects the couple's desire to focus on and maintain their relationship. These long-term lovers also showed increased activity in areas of the brain rich in opioid and serotonin, which we don't see in those newly in love. These regions have the capacity to modulate anxiety and pain and are central brain targets for the treatment of anxiety, OCD and depression. This suggests that as we move through a loving relationship our brains help us to experience a greater amount of calm and happiness, too.[1]

This might be part of the reason why getting married, or finding a long-term life partner, can have such an impact on your longevity. Some studies suggest it can increase your lifespan by up to seven years. As well as boosting happiness, a life-long partner is also linked with lower levels of stress and risk of depression, less social isolation and a lower incidence of heart disease. It can also protect against cancer. It doesn't stop you getting the disease, but it can affect the outcome – unmarried individuals are more likely to have advanced disease by the time they are diagnosed than married people.

So stable, loving relationships are good for us. How do we start?

HOW TO WOO A LOVER

As I mentioned, I've been on my fair share of dates. When online dating was a fairly new and somewhat embarrassing concept, a friend at work persuaded me to join up – as was common back then – to a web-based dating tool. Now, people are more likely to use apps like Tinder and Hinge. Statistics seem to suggest online dating is a reasonably good bet, too. Around a third of American adults have used online dating sites or apps, and 12 per cent say they have married or been in a committed relationship with someone they met through using these sites.

If you use online dating sites there a few things worth bearing in mind. Firstly, don't sell yourself short: most people pursue partners who are roughly 25 per cent more desirable than they are themselves. And while you might be tempted to agonise over the content of your messages, it's probably not worth it. By my final online date, I had got thoroughly bored of writing long-winded messages, getting excited by weeks of witty repartee, and then knowing instantly that I didn't fancy the guy in person. I responded to user Superafin's first message to me with the reply: 'Hi, I'm kind of over the long back and forths, do you wanna just meet up next Tuesday?'

It turns out that was probably the best way to go. A 2018 study on online dating messages revealed that while women get a slightly higher reply rate when they write longer messages, men see the opposite effect. Overall, the variation in pay-off for different writing strategies is tiny, suggesting that a lot of effort put into writing long, positive messages may be wasted.[2]

When it comes to meeting up with your date – or if you're going about things in an old-school way, approaching an attractive man or woman in a bar – you might be worried about the perfect chat-up line. In reality, our body gives away a great deal before we open our mouths. When we meet a stranger, it's estimated that their impression of you is based 55 per cent on your appearance and body language, 38 per cent on your style of speaking and a mere 7 per cent on what you actually say.

To give the best first impression, adopt an open posture with no folded arms. As we saw in Chapter 4, mirroring the other person's posture can help to create a feeling of affinity. You can also indulge in a little 'gestural dance', synchronising your gestures and body movements, such as taking a sip of your drink at the same time as your potential mate's. Most people aren't conscious of being mirrored in this way, but evaluate those who do it more favourably.

Any flirt knows that making eye contact is an emotionally loaded act. But psychologists can show just how powerful it can be. When pairs of strangers were asked to gaze into each other's eyes, it was perhaps not surprising that their feelings of closeness and attraction rocketed compared with, say, gazing at each other's hands. More surprising is that one couple who met during such an experiment ended up getting married.

Neuroscientists measuring brain activity during this kind of experiment see that meeting another person's gaze activates regions of the brain associated with reward and pleasure. But again, as we've learned before, make sure you don't hold the gaze too long. If it's not reciprocated or you forget to blink you risk making the other person very uncomfortable.

When I was researching this topic, my thoughts kept returning to my second date with 'Superafin' – my husband. Our first meeting had been great – we both had an immediate physical attraction to one another and our conversation flowed easily – but the follow-up wasn't going well. He was hungover, and the conversation was stilted. A third date was looking less and less likely. Then, as we walked along the river to say our good-byes, an elderly woman suddenly collapsed at my feet. She started shaking uncontrollably, clearly having a seizure. Both Alex and I reacted quickly. I'd had basic first aid training and grabbed my jumper to protect her head while the fit took its course, and then placed her in the recovery position once it had ended. Alex phoned 999 and shouted to the crowd to see if there were any medics around. Being in central London, a first responder came over within minutes and took over the woman's care.

It was all pretty dramatic, and I'm sure it changed the course of my life. I'll never know for certain, but I'm convinced it was the only reason we ended up agreeing to meet up again – a much better third date that led our relationship to where it is today. That anecdotal experience tallies with numerous studies showing that a dramatic setting or meeting someone when physiologically aroused increases the chance of having romantic feelings towards them.

That's because of a strong connection in the brain between anxiety, arousal and attraction. In a classic 'shaky bridge' study, carried out by psychologists in the 1970s, men who met a woman on a high, rickety bridge found the encounter sexier and more romantic than those who met a woman on a low, stable one. A visit to a funfair works wonders too. Photos of members of the opposite sex are more attractive to people who have just got off a roller coaster, compared with those waiting to get on. For the same reason, think about your choice of a movie date: couples are more loved-up after watching a suspense-filled thriller than a calmer film. Why? No one is sure, but the adrenaline rush from danger, panic or excitement might be misattributed in the brain to the thrill of attraction.

Rocky bridges are probably best left alone, but there are a few other ways to get your partner's heart racing. Just as the peacock's prodigious plumage suggests its owner is physically fit, so too might your choice of clothes. Around 2010, several studies suggested that wearing red made you more attractive to the opposite sex. (Other studies showed that sports teams or individual athletes were more likely to win their events when dressed in red.) Various explanations were given. The colour red has symbolic and cultural meanings that could evoke an unconscious emotional reaction. Colour also carries biological signals – reddening of the skin could potentially suggest sexual receptivity in females, for instance.

However, three attempts to replicate these studies in 2016 failed to show any effect of wearing red on dates, leading the authors to conclude that if there is an effect it's likely to be small. Your best bet is to wear whatever makes you feel most confident. It sounds obvious, but this small mental boost can really shine through, in women at least. When asked to rank mugshots of women, men consistently chose pictures of women who were wearing their favourite outfits, despite the fact that the women were asked to keep their expressions neutral – and their clothes were not visible. The way the women felt about their appearance was apparent in their faces even though they were not consciously aware of showing it.

One last tip: when you're trying to make conversation with the apple of your eye, use lots of short, snappy words of encouragement – like 'go on', 'OK' and 'I see'. In real-world tests, individuals who do this seem to be rated as more attractive by their date. It's not surprising really, given that it makes you feel listened to and interesting, but an easy thing to forget when you're nervous or self-conscious. If you need more on this, turn back to Chapter 4 on making friends.

HOW TO IDENTIFY 'THE ONE'

By now you might be a pro at dating. But how do you know when to stop? Selecting a partner can be one of the most crucial decisions of our lives and we devote a huge amount of time and energy to it. Our appetite for a relationship fuels an industry of matchmaking and online dating services. Yet we're often not satisfied. Despite many people successfully finding love on dating sites, 45 per cent of Americans also say the experience left them feeling more frustrated than hopeful.

We seem as much in the dark as ever about who is a suitable match for us. This isn't particularly surprising – the way we go about choosing a partner has been a source of much mystery for scientists as well. Mate selection, as they like to call it, is a highly complex process. We're consciously aware of only part of it, and the rest either operates outside our awareness or is inherently unpredictable. It's no wonder we often refer to love as being about some ineffable chemistry.

That said, a few things might help point you in the right direction. There are some criteria that, in the main, we all find attractive. Heterosexual men tend to like women with features that suggest youth and fertility, such as a low waist-to-hip ratio, full lips and soft facial features, while heterosexual women have strong preferences for taut bodies, broad shoulders, clear skin and defined, masculine facial features, all of which may indicate sexual potency and good genes. We also know that women are attracted to men who look as if they have wealth, or the ability to acquire it, and that both men and women strongly value intelligence in a mate. While it all sounds pretty shallow, preferences for these qualities – beauty, brains and resources – are universal. The George Clooneys and Angelina Jolies of the world are sex symbols for predictable biological reasons.

Of course, we don't all fall in love with people like these. It's probable that part of the reason why love evolved is to bond us for cooperative child-rearing, and also to assist us in choosing, so that we don't waste time and energy falling for someone who is unattainable. Instead, people tend to fall for others who, on attractiveness, intelligence and status, are of a similar 'ranking' to themselves.

There are also less obvious rules of attraction. Geneticists have shown that each of us will be attracted to people who possess a particular set of genes, known as the major histocompatibility complex (MHC), which play a critical role in our ability to fight pathogens. Mates with dissimilar MHC genes produce healthier offspring with broader immune systems. And the evidence shows that we are inclined to choose people who suit us in this way: couples tend to be less similar in their MHC than if they had been paired randomly.

Despite decades of research, it is still not particularly clear how we identify people with different MHCs to our own. It may be about smell – people tend to rate the scent of T-shirts worn by others with dissimilar MHCs as more attractive. Perhaps this is the sexual 'chemistry' we often refer to.

The message seems to be to trust your instincts – with one alarming exception. Women who take hormonal contraceptives tend to prefer men whose MHC genes are similar to their own. It means that women on the pill risk choosing a partner who is not genetically suitable. There's no study I'm aware of that has analysed whether couples may be significantly more likely to break up after a woman comes off hormonal contraception than at other times, but I took this piece of advice to heart while living with my now husband. I had written about MHCs so much that I needed to make sure. I came off hormonal contraception after Alex proposed to me, just to make sure my feelings hadn't been masked by chemicals.

Attraction also fluctuates over the menstrual cycle. Men

evaluate women's scents as more attractive when they are near ovulation, and are more loving towards their partners as ovulation approaches. Women's preferences for certain male scents and other male features also change over their cycle. Near ovulation, they prefer masculine traits; at other phases they prefer less sexiness and more stability.

Having sex can also complicate the way you perceive a potential partner. After sex, the brain releases oxytocin, which results in that warm, companionable feeling of love and the creation of the social bond that facilitates cooperative child-rearing. That is great in certain situations. But it also means that sex on a whim can lead to temporary feelings of compassion for a person who is entirely wrong for you.

So we're still left with the big question: if the roads to love are so varied, how do we decide on one particular partner? One way might be to leave it up to maths. Researchers at the University of New Mexico used a computer simulation to determine how a person might best choose from a number of potential partners. They set it up so that the person first assessed a proportion of the options before them to decide what was the best they could aspire to in terms of attractiveness, and then went for the next person they came across who met their aspirations out of those they hadn't already encountered.

The researchers found that the optimum proportion of possible mates to examine before setting your aspirations and making your choice is a mere 9 per cent. So at a party with a hundred possible mates, it's best to study only the first nine you randomly encounter before you choose. Examining fewer means you won't have enough information to make a good choice, while examining more makes it more likely you'll pass the best mate by.

Of course these models underestimate the complexity of

real-mate choice, but perhaps the lesson is not to search for too long – lest we miss out on someone who, given a little more of your attention, might turn out to be perfect.

HOW TO HOLD ON TO LOVE

In 1838, the naturalist Charles Darwin asked himself a question. Should he propose to his cousin, Emma Wedgwood? To help him make up his mind, he scribbled a list of pros and cons on the back of a letter from a friend. The pros included a wife being, in terms of companionship, 'better than a dog'. He considered the charms of female chit-chat, and the possibility of children. He also thought these things were 'a terrible loss of time', and would result in a lack of freedom to go where he liked, potential quarrelling, as well as making him fat and idle.

While your own pros and cons may differ, there's no doubt that moving in with someone, sharing bills, planning a wedding and so on are a big deal. When it comes to spending the rest of your life with someone, how do you make sure it lasts the distance?

Social scientists first started observing marriage seriously in the 1970s in response to a crisis: an unprecedented divorce rate. One of those researchers is John Gottman, a man once described as having 'the spirit of a scientist and the soul of a romantic'. Now pushing eighty, Gottman has been voted one of America's top ten most influential psychotherapists. He has authored hundreds of papers on relationships, marriage stability and divorce prediction, and runs the Gottman Institute, a research organisation whose work earned him the title of 'The Einstein of Love'.

Some of Gottman's seminal work separated couples into 'masters' and 'disasters': those who had been happily married for more than six years, and those who were divorced or unhappily married. He wanted to know whether those who end up 'happily ever after' behaved in fundamentally different ways to people whose relationship was more toxic.

For instance, his team observed 130 newly-wed, heterosexual couples as they went about their lives. Gottman came to an important conclusion: that all individuals would make what he called 'bids'. The husband might comment on a car parked outside the house, for instance. The wife has a choice. She can either engage positively with him in her response – perhaps asking what car he would buy if he had unlimited funds, or she can answer minimally or not at all, preferring to engage with whatever she is doing at the time. Gottman called the former 'turning-toward bids' and the later 'turning-away bids'. In follow-up experiments six years down the line, couples who had divorced had turn-toward bids 33 per cent of the time, compared with those who were still together who responded positively 87 per cent of the time.

Gottman performed hundreds of such experiments. Often, they demonstrate obvious acts of kindness between couples that work out and clear acts of disrespect between those that don't, but seeing these behaviours spelled out can often be enlightening. Just knowing about the 'bid' experiment, for instance, often makes me conscious of how I respond to my husband.

Back in 2006, Gottman spoke to *New Scientist* and summarised the questions you should ask yourself at the start of a relationship, which if you can answer 'Yes' to them suggests you're on the right track. Are you getting treated with love, affection and respect? Do you feel there is a basis in terms of nurturing, support and affection? Do you really like spending time with this person, so that the time sort of flows like wine? Is it easy to be together? Do you like yourself when you are with this person?

He says that in his experience, as people get closer, the other thing to look for is the feeling that you can create a sense of shared purpose and meaning and values. This is in line with what we see when we study the brains of people in long-term relationships – regions are activated that are associated with shared goals. For Gottman, answering yes to these questions is a good signal that you should move forward together.

Over his long career, Gottman has also come across people who are naturally good at bonding with others. He says they have a habit of mind where they are looking for things to appreciate. They are looking for things to say 'Thank you' about. At the other extreme, those who are bad at long-term relationships focus on their partner's mistakes. They are scanning for what their partner is doing wrong. Putting this research together, he has come up with four things that are more corrosive to a relationship than anything else. He calls them the 'four horsemen of the apocalypse': contempt, superiority, criticism and stone-walling.

The most negative is contempt: direct insults and sarcasm. An air of superiority, feeling that you're better than your partner is by itself the best predictor of divorce, he says. Criticism is another sign of a relationship going nowhere, as is defensiveness, such as if you respond to a complaint with righteous indignation or as if you are an innocent victim. If you behave this way, you don't take any responsibility for the problem. Meanwhile stone-wallers withdraw emotionally from an interaction and don't give the usual non-verbal signals to the speaker. They look away or down and stop responding.

One final secret of success is not about how you fight, but

how you make up. Those in successful relationships will notice that some aspect is not going well and will attempt to repair the situation. But rather than there being a perfect way to apologise, the key is in accepting that apology. Individuals who accept their partner's attempts at repair, regardless of how bad the apology is, are more likely to stay together than those who don't acknowledge the effort.

HOW TO MEND
A BROKEN HEART

It's a sad fact that sometimes, no matter how much we want something to work out, it doesn't. Just like most of you out there, a few of my relationships have ended with obsessive texting and crying under a blanket. When you're in the midst of heart-ache, there's nothing more earth-shattering than the physical and emotional pain of unrequited love.

As we've seen earlier, being in love is a lot like addiction – they both activate our reward system – so it's not surprising that when we get dumped we find it difficult to give up that habitual compulsion to see, hear or touch the object of our desires. When anthropologist Helen Fisher studied people who had recently been left in a break-up, she found activity in the brain that resembled the cravings seen in gambling or substance abuse. And don't let anyone tell you that the pain you're feeling isn't real – her team also found activity in areas of the brain responsible for physical pain and the associated anxiety.

In 2019, scientists came up with one possible, albeit extreme, solution to a broken heart: propofol, a commonly used sedative. Researchers in Spain found that the drug made upsetting memories less vivid. When we retrieve a memory, there is a short window afterwards in which it is possible to modify that memory. The team wondered whether propofol might affect this, so asked fifty volunteers to memorise two upsetting stories from slide shows, a week before they were due to be sedated for surgery.

Immediately before being sedated, each volunteer was shown the first slide from one of the stories and asked several questions to reactivate their memory of the tale. Straight after their procedure, half were tested on how well they recollected the stories. These volunteers remembered both stories equally well. The rest of the volunteers were tested twenty-four hours later, which gave propofol time to take effect: these participants were 12 per cent worse at remembering the emotional parts of the reactivated story compared with the non-reactivated one. That's probably because the brain circuitry involved in emotional memories is quite sensitive to anaesthetics.[3]

The goal of this experiment was to help lessen the emotional impact of traumatic memories associated with post-traumatic stress disorder (PTSD). There is a risk to taking sedatives of course, so any use would have to balance the drawbacks and benefits, but the researchers involved suggested that there may be people for whom heartbreak is so distressing that it warrants such measures.

Others have suggested that a cure for love already exists in the form of selective serotonin re-uptake inhibitors, or SSRIs. A common side effect of these widely prescribed antidepressants is a blunting of emotional responses, including towards loved ones, so it has been proposed that they might be helpful in treating the anguish that can follow a failed relationship.

And while you might be familiar with taking paracetamol for a painful headache, have you ever considered that it might help get you over your emotional heartache, too? Physical pain and social pain, such as that caused by a rejection, are controlled by overlapping neural systems. Paracetamol acts centrally, easing pain by blocking chemical messengers in the brain, so it makes sense that it might also help cure social pain. In studies, paracetamol taken daily for three weeks helps people experience significantly fewer hurt feelings than those who take a placebo. Likewise, when people take paracetamol for three weeks before being made to feel socially excluded, scans of their brain showed less activity in response to the social pain than those who take a placebo.[4] Further research shows that this effect is increased if you also spend time each day thinking about forgiving the person who is responsible for your pain.

One word of warning: recent research suggests that taking paracetamol also reduces empathy for other people's suffering. Since we rely on empathy to be decent human beings, it raises questions about the impact of increased paracetamol consumption. That's before we even consider the risks associated with taking any drug, even such an everyday one as paracetamol.

And of course, the whole idea of taking drugs to cure the pain of lost love brings up a range of red flags. Might it one day be possible for someone to use such drugs to intentionally sever a relationship? Could tampering with fraught memories also interfere with those we want to keep? Would it make it harder to deal with future events that could have been avoided had we more clearly remembered similar experiences in the past?

All in all, if you're suffering from heartache it might be safer sticking to a simpler approach. Just like any addict, you need to cut off your supply. No calls or texts or spending time staring at old pictures. Then replace your fix with something else that gives you a burst of the feel-good hormones dopamine or oxytocin. Exercise will ramp up your dopamine and bodily contact and social interaction can raise oxytocin.

In the end, time does heal: we see that brain areas responsible for feelings of attachment diminish in activity in response to thoughts of a lost love over several months. It's something my mum always said, and it turns out to be true. When it comes to a broken heart, a little time – and possibly a new lover – truly are the best antidote.

HOW TO CREATE THE PERFECT FAMILY

On that first date with my husband, I sent a rather embarrassing text to my housemate. It said: 'I'm on a date with the future father of my children'. Four years later at our wedding, humiliatingly, that text made a reappearance during the speeches. As prescient as it may have been, trying to make that text come true was a different matter. In the end, it took five years, five rounds of IVF and three donated eggs from two anonymous women to help me conceive my two children. Because of this, I am particularly interested in modern families, and how well they work in all their different guises. Decades ago, the debate about the perfect family structure focused on the merits of a mother, father and their biological children versus the extended family. This discussion has changed dramatically.

Now, more than half the children in the UK and United States are being brought up outside a nuclear family, and not because of any great revival of the extended family. Thanks to the rise in reproductive technologies such as egg and sperm donation and permissive social attitudes, today's families are more diverse than ever. What does that mean for our relationships and our children?

To some people, any deviation from the norm is unacceptable – especially if it involves same-sex relationships. In the run-up to Australia's national survey on gay marriage, shocking posters appeared in Melbourne claiming that 92 per cent of children raised by same-sex parents are abused, 51 per cent have depression and 72 per cent are obese.

The statistics on the posters do come from real studies, albeit ones that have been debunked. The scientific consensus is the exact opposite. The near-unanimous conclusion from a body of long-running studies on the emotional and psychological health of children brought up in non-traditional families of all kinds is that there is nothing to worry about, over and above regular concerns about child welfare within any family. Some studies even suggest that welfare is higher in families that have used IVF or adoption, simply because of the emotional investment made to start them.

There are questions concerning identity and relatedness. Here the research is a little less clear, but the emerging message is the same: there is no extra risk to children brought into the world in this way, as long as you are open and truthful with them about their background from day one. In thirty years' time the debate about alternative family structures will hopefully appear as quaint as the one over nuclear and extended families. Perhaps the conversation will have moved on to another question some of you may have considered: do we need to limit ourselves to just one partner?

The lifelong commitment of two people to one another is the fairy-tale ending of Western society. But monogamy is a relatively modern development and not always a sure path to happiness. One idea for how monogamy came to dominate our once polygamous societies is that as we evolved larger brains looking after babies required more effort and food. The children of men who were spread across too many families were less likely to survive. A recent analysis found that, from hunter-gatherers to industrial societies, the greater the father's investment in his offspring the more monogamous the society.

The invention of weapons may have levelled the playing field, because dominant men were no longer able to fend off competitors who were weaker but armed. That aligns with another idea: that monogamy helped social stability. If a few men monopolise all the women, that leaves a lot of disgruntled bystanders. Monogamy was a social trade-off, whereby powerful polygamous men gave up their harems in return for a degree of social peace. Religion also played a part, making polygyny and extra-marital sex increasingly socially unacceptable.

Norms aren't so rigid today. The erosion of religious values, the development of hormonal contraception and the rupture of taboos around sex and divorce means many of us are serial monogamists, moving from one long-term relationship to the next. Even so, we're pretty bad at staying true. In a 2015 UK poll, 1 in 5 people admitted to cheating on a partner.[5] Perhaps that's why some people have abandoned the idea of monogamy altogether. A 2016 survey found that 20 per cent of single people in the United States have had consensual non-monogamous relationships, where people have multiple sexual partners openly.

Can these relationships really work, given the hurt and outrage infidelity routinely causes? Where devout monogamy is expected, it's no surprise that infidelity spurs negative feelings. But it is often the betrayal of trust that proves most harmful. Consensual non-monogamous relationships at least cut out the deceit.

When psychologist Terri Conley at the University of Michigan and her colleagues compared people in traditional monogamous and open relationships, they found no significant differences in reported relationship satisfaction, commitment or passionate love. What's more, those in open relationships reported less jealousy and higher levels of trust. They're not doing so much better than monogamous couples that everyone should make the switch, but they are doing OK.

So at the very least there is room in our society for other types of relationship. The systems that shape how we select our partners are flexible, and changeable. At some point, it became ingrained that monogamy is what we do. But that doesn't mean it is the only way, or the best one. The final consensus seems clear. No matter what path you take to romance, or how you structure the family unit, the kids – and the adults – will do just fine, as long as one condition is fulfilled: that there is mutual love and respect.

TOP TIPS FOR FINDING LOVE

🔧 On a first date, use lots of short, snappy words of encouragement – like 'go on', 'OK' and 'I see'. In real-world tests, individuals who do this seem to be rated as more attractive by their date.

🔧 Ask yourself a few choice questions about any potential long-term partner: Are you getting treated with love, affection and respect? Do you feel there is a basis in terms of nurturing, support and affection? Is it easy to be together? Do you like yourself when you are with this person? If the answer is yes, you're more likely to stay together for the long haul.

🔧 Be on the look-out for the 'four horsemen of the apocalypse': contempt, superiority, criticism and stonewalling. These four behaviours are the most predictive sign that a relationship will break down.

🔧 If you experience heartache, time truly is the best healer: we see the pain associated with break-ups diminish in the brain over several months. But in the meantime, take some exercise and seek the comfort of another's touch – this will ramp up feel-good hormones like dopamine and oxytocin, giving you a mental and physical boost.

🔧 Forget tradition. No matter how alternative your family structure, studies on the whole show that the kids and the parents have as much chance of being happy and healthy as those in more traditional relationships.

6

HOW TO LIVE HEALTHIER FOR LONGER

MY HUSBAND'S grandmother, Margaret, is ninety-two years old. We recently went round to her house for lunch. She cooked the food, opened a bottle of wine and insisted that she lay the table. She got out some sixty-year-old toys for my daughter to play with, and sat and reminisced with us about how my mother-in-law would play with the same wind-up mouse for hours on end.

As one of the UK's first female doctors, it's probably not surprising that she has stayed physically and mentally healthy for such a long time. She claims it's 'keeping busy' that has done it. Indeed much of her time is spent with book clubs, weekly mah-jong with friends and enjoying her beautiful garden with family.

In India, 116-year-old Battle Lamichhane claims the secret to long life is smoking thirty cigarettes a day for ninety-five years. The oldest person ever, the Frenchwoman Jeanne Louise Calment, who died at the grand age of 122 – at least officially, as doubts have been raised over her identity – was another smoker. She also enjoyed a huge amount of chocolate, ate spicy foods and drank a small amount of port with every meal. At 112, American Richard Overton said it was cigars and a Bourbon in his morning coffee that was key to his long life.

The physical side of health, and how to live a long life in good trim, is not my primary focus in this book. If anyone wants to know in fine detail what the evidence says about what diet or exercise regimes are effective, the secrets (or not) to supplements and the ideal way to lose weight, check out the companion book to this one, *This Book Could Save Your Life*, by my *New Scientist* colleague Graham Lawton.

Nevertheless, there is a lot of truth in the old adage 'a healthy mind in a healthy body'. It would be a huge omission for any book aiming to show you how to lead a happy, fulfilled, successful life if it ignored physical health and the interplay between mind and body. It's this interplay that will be my focus in the coming chapter. It may seem obvious that a healthy body leads to a healthy mind, and we'll investigate some of the specific ways the two are linked. But it's perhaps less obvious that it works the other way around too. A positive mindset can trigger changes that make you fitter, slimmer, more energetic and less stressed. It can help you overcome laziness, boost motivation – and ultimately help you stay fitter for longer.

HOW TO UNDERSTAND 'HEALTHSPAN'

There's no getting around the fact that every day you're getting a little bit older. That's mainly due to the breakdown of processes designed to make us a little younger every day. Everyday life puts a huge strain on the cells and organs of our bodies, and evolution has ensured that we have a suite of repair processes designed to rejuvenate and replace damaged cells. But naturally enough, they only work fully in the younger stages of our life before we pass our genes on.

As we pass the sort of age where we naturally reproduce, these repair mechanisms begin to break down. Organs and tissues clog up with clumps of protein and other detritus, genetic mutations accumulate and chromosomes start to unravel. Some cells become cancerous, while others that would earlier have just stopped working, awaiting removal by white blood cells, just sit there. Some scientists call them 'zombies', because they don't do anything useful except cause havoc, pumping out inflammatory proteins that cause damage to the surrounding tissue.

The result is that our immune systems weaken, our muscles lose mass, fat builds up around our internal organs and we find ourselves with low level inflammation and a lack of cells that produce much needed energy. All of this leads to the inevitable diseases of age, such as atherosclerosis, Alzheimer's, cataracts, type 2 diabetes and so on. Unpleasant it may sound, but this is just life. But ageing does not hit us equally – and nor is a long life necessarily a healthy or happy one. Margaret, my husband's grandmother whom I mentioned just now, enjoyed a vast number of years in relatively good health. Meanwhile one of my own grandmothers died at ninety-seven, but having spent decades blind, physically frail and suffering from dementia.

Over the past century, the average human lifespan has increased from around thirty years to now above seventy, and eighty or more in many wealthy countries. What scientists are now more concerned about, however, is our 'healthspan', where the increase has been far less dramatic. This refers to the number of years we have to live that are at an acceptable level of vitality, not burdened by the chronic physical or mental diseases that often accompany old age. A 'quality-of-life-span' if you will, which is a desirable goal for us all.

Extending healthspan means stopping or slowing some or all of those processes that I described in the preceding paragraphs. In the very near future, popping a pill might be an option. Drugs called senolytics go on a kind of seek-and-destroy mission around the body, clearing out the knackered cells that cause ageing. They are no longer a pipe dream: trials of these drugs are going on and there's talk of them being in the clinic in under five years.

Until then, there are other promising ways to increase health-span. One way that looks good in animals, but is by no means confirmed in humans, is caloric restriction – fasting. It seems to affect a series of cellular processes – a 'pathway', in the jargon – called mTOR, which is thought to have evolved to help us survive periods of starvation. If you eat, this pathway is switched on, telling cells to divide and grow. If you don't eat, it switches off. This triggers protective pathways, which include those that scavenge dysfunctional organelles and molecules for energy. This creates a waste disposal system for damaged cells and therefore helps remove the zombies that build in tissues and organs and slows the ageing process.

Early evidence suggests that mTOR gets stuck in the 'on' position as we age, so caloric restriction may help to switch it back off. The evidence is still not clear-cut in humans and nobody should be starving themselves just yet. Given our relatively long lifespans, it will take several years and several long-term studies to really get to grips with its effects, although accumulating data from several clinical trials indicate that we're on the right track: periodic fasting results in some of the same physiological adaptations in humans that have been shown to improve health and slow the damage of ageing in animals. There doesn't seem to be any harm in lowering your calorie consumption one or two days a week.

Anti-ageing nutritional supplements such as Rejuvant and Basis that are now trickling on to the market are designed to switch off mTOR and have been shown to work in mice. One word of caution: although they've been proven to be safe for human consumption, no long-term studies have confirmed their effect on ageing yet.

Popping a pill or starving yourself might not be your thing. And meanwhile there is one miracle cure freely available to all that's guaranteed to increase your healthspan and slow the ageing processes in your body and your brain – in part because it too helps trigger processes that clear dying cells. You just have to select the course of treatment that's appropriate to you. I'm talking, of course, about exercise.

HOW TO CHOOSE THE BEST WORKOUT FOR YOUR BODY . . .

In the past fifteen years I have signed up to a gym on four separate occasions, have been a regular visitor to my council pool, and have paid for a co-working lounge because it had a free weights room. I have given money to an app that gives me access to hundreds of fitness classes in London. I've also been an on-and-off bootcamp devotee, taken salsa lessons, learned circus skills, tumbling and tight-rope walking, worked out in the dark with glow sticks, spent hours performing Pilates in my community greenhouse, somehow managed to complete a marathon, sweated through a yoga session in a heated pod and even briefly – very briefly – considered naked yoga.

There's a reason for my (sometimes misplaced) enthusiasm. Exercise is pretty much a miracle-worker if your goal is to live a long, healthy life. It can fend off cancer, obesity, diabetes, depression and heart attacks. It prevents more premature deaths than any known medical drug, and when used correctly it has zero side effects. But while we all know that exercise is good for us, it's only recently that we've realised just how good, and how different types of exercise are best for certain aspects of your health.

I can never quite pin down the right exercise for me. I'm not alone: apparently Brits spend more than £500 million a year on gym memberships that they don't use. You might think that any exercise is better than nothing, and you'd be right. But it's worth giving a bit of thought to what you're doing to get your body moving.

Most of us are familiar with the basic menu of exercise options and how some of them can shape and benefit different areas of the body. Pumping iron sculpts your biceps, running helps strengthen your heart. When it comes to choosing the right workout, however, it's helpful to go back to basics, because even there you could be missing a trick or two.

For years, aerobic exercise has been seen as the holy grail of fitness – and for good reason. Aerobic fitness boils down to how effective your body is at delivering oxygen to muscle cells. If you increase your aerobic fitness, you are growing your muscle fibres and supplying them with a more efficient network of blood vessels. You're also increasing the size and number of your mitochondria, the powerhouses of cells.

As your heart pumps harder to provide your body with energy, the stress imposed on arteries promotes the production of nitric oxide, a muscle relaxant that keeps your blood vessels stretchy and helps to repair any damage. As blood courses through your body, it flushes out fatty deposits in the walls of blood vessels that can clog them up, causing heart attack and stroke.

Aerobic exercise also protects us from diabetes by reducing the damage caused by dangerous fats and helping the body remove glucose, which can contribute to the disease. It also reduces the risk of cancer. It's not entirely clear how, although it might be related to reducing obesity, which is a risk factor for the disease.

On top of all that, aerobic movement stops us being lazy, which can be harmful. People who sit regularly for one to two hours at a stretch have a significantly higher risk of early death than those who spend the same amount of time sitting, but who get up and move around every half-hour or so. Our laziness produces a complex cascade of health problems. Unused muscles shrink, and shift the way they work, burning less fat. Unburned lipids accumulate in the blood, and fat gathers in muscles, the liver and the colon – all places where you don't want it to gather.

So which type of aerobic exercise should you choose? First, give a large dose of scepticism to media coverage that suggests something amazing about a certain kind of exercise. It's largely driven by small-scale studies, in which an unusual finding from a small group of people gets far more attention than the large pile of studies that counter it. One study that shows running makes you fatter, for instance – it doesn't – is going to get into the newspapers and around the web quicker than ten that show it makes you leaner.

In the same breath, try to ignore celebrity workout regimes that promise to change your physique and fitness in a matter of weeks. If it sounds too good to be true, it probably is. It's unlikely that a celebrity has found the secret to a six-pack faster than decades of sports scientists.

Most government guidelines recommend avoiding long periods of being stationary, and advise adults to get at least 150 minutes of moderate aerobic activities a week. These include running, brisk walking, ballroom dancing, gardening – anything that gets your heart pumping and you breathing faster. That's all fine and good – but aerobic fitness is far from being the whole story. Another kind of workout is just as important for our health, if not more so, and it's something we can all do from the comfort of our homes without any equipment: strength training.

Our muscle strength is at its peak during our twenties through to our mid-thirties, then slowly declines. We start to lose up to 5 per cent of our muscle mass each decade, and this accelerates at seventy. Eventually, it can drop so much that we can't get out of chairs or climb stairs. But strength training isn't just for older people. There are unexpected health benefits from building muscle for all adults that go way beyond simply being strong. Its importance is so great that the UK government's latest physical activity guidelines emphasise muscle strengthening over aerobic workouts, suggesting we all do at least two sessions of strengthening activities per week.

The best evidence for its benefits comes from studies of the exercise habits of large numbers of people. One showed that lifting weights for less than an hour a week reduces the risk of heart attack and stroke by up to 70 per cent, independent of any aerobic training. Another study of 100,000 women found that those who did at least an hour a week of strength training significantly lowered their risk of type 2 diabetes. And people with better grip strength – a proxy for overall muscle strength – have a lower risk of cardiovascular disease and cancer and are at reduced risk of premature death by any cause.

How do strong muscles equal better health? The most obvious reason is that they help prevent the debilitating effects of wobbles, falls and problems moving, increasing well-being in the process. But muscle also plays an important role in regulating the body's glucose levels. With the help of insulin, muscles soak up glucose from the blood and store it in the form of glycogen. Bigger muscles mean a bigger sink for glucose and a higher number of cells that transport and clear glucose from the body, which all helps ward off type 2 diabetes, in which blood glucose levels become too high. The better survival rates for people with cancer are probably because the disease decreases muscle mass, so it is helpful to have a bigger resource to start with to keep the body going for longer.

You may have heard that increasing your muscle mass significantly increases your basal metabolic rate – the amount of energy your body consumes when at rest. This is a kind of half-myth. While pumping iron isn't going to suddenly make you burn calories while you sleep, you will continue to burn calories for a short time after the exercise is over. Bigger muscles require more energy to fuel their tissue maintenance, so simply having more muscle mass does use more calories. But this amounts to only about ten to fifteen calories per kilogram of muscle per day. A regular weight programme will probably add only two kilograms of muscle, which means your bigger muscles burn fewer calories than you'll find in a single Jaffa Cake.

There's another effect to consider, however. Lifting weights causes tiny tears in your tissue that require a relatively large amount of energy to remodel. This increase in energy demand can last three days after a workout. Let's say I fit in two twenty-minute resistance-training workouts a week. Each session requires about 200 extra calories to perform, but over the next three days I will use up to 100 extra calories a day to help repair my muscles. Over the month, my two workouts a week could have consumed up to a whopping 5,000 extra calories – without even leaving the house.

All of this helps if you want to decrease body fat, which

besides helping you to look more toned, is also a factor associated with lower cholesterol, lower blood pressure, and improved insulin sensitivity and glucose control, all of which contribute to a decreased risk of type 2 diabetes and cardiovascular disease.

But strength training really trumps aerobic exercise with its effect on bone. Our bones start to degrade as we age, losing mass and making us more prone to fractures. Aerobic exercise is beneficial to a lot of systems in the body, but there is little evidence that it protects us from bone loss. Our bones are in a constant flux of being broken down by cells called osteoclasts and being built up again with osteoblasts. Strength training places stress on the bones, triggering the activity of osteoblasts and inhibiting osteoclasts, helping us to maintain, and even build, denser bones. This significantly lowers the risk of osteoporosis, which causes around 1.7 million hip fractures globally every year.

So what's the best way to reap the benefits? The type of strength exercises a person can do will differ wildly depending on their age and circumstances. That said, advice from the American College of Sports Medicine couldn't be simpler: it says that adults should perform strength exercises on all major muscle groups – legs, hips, back, abdomen, chest, shoulders and arms – at least twice a week. That advice comes from evidence that your first workout of the week will give you the most benefit compared with nothing at all. Your second workout will give a bit more benefit, as will the third, but then the results plateau.

But don't stress yourself with the detail. If you exercise a particular group of muscles until it's tired, it doesn't really matter how heavy the weight is or how many times you lift it. The benefits for a non-athlete are broadly the same whether you lift a light weight twenty times, or a heavy weight five times. Little things squeezed into your everyday routine can make a big difference, without the need for any equipment – press-ups against the kitchen worktop, squats in front of the TV, or just lifting your children all count.

There's just one rule: whatever you do, make sure it wears you out. If you do whatever strength exercise gets you tired in a reasonable amount of time, you'll probably get the same benefit to your health as if you were following a highly specific training routine.

. . . AND YOUR BRAIN

Things get a little more nuanced when it comes to thinking about what kind of exercise is good for your brain. It's only relatively recently that research has begun to highlight the profound effects certain types of exercise can have on your mental faculties and well-being. If you were to peer inside the heads of people who like to keep active, you'd see that different exercises strengthen, sculpt and shape the brain just as they do the body, boosting creativity, helping you focus for an exam, minimising your stress levels and even curbing your cravings.

We first got a hint that exercise affects the brain from studies in the 1960s, but its importance was more appreciated in the 1990s, when researchers discovered that exercise cultivated the growth of new neurons in mice. As a result of working out mice showed improvements in memory that allowed them to navigate mazes better.

Similar indications soon came from humans, too. Older adults who did aerobic exercise three times a week for a year grew larger hippocampi (the hippocampus is a brain area intimately involved in remembering), and performed better in tests of memory. Several studies show that people with a better grip strength also score higher on tests of attention and reaction time, as well as on assessments of verbal and spatial abilities.

Working out whether one kind of exercise is better than another in the context of brain health is a little harder. In one study researchers compared the effects of aerobic exercise and strength training in eighty-six women with mild cognitive impairment. One group lifted weights twice a week for an hour, while the other went for brisk walks quick enough that talking required effort. A control group just stretched for the hour.

After six months of this, both walking and lifting weights had a positive effect on spatial memory, the ability to remember surroundings and sense of place. On top of that, each exercise had unique benefits. The group that lifted weights saw significant improvements in their executive function, which encompasses complex thought processes, such as reasoning, planning, problem-solving and multitasking. They also performed better in tests of associative memory, which is used for things like linking someone's name to their face. The aerobic-exercise group saw improvements in their verbal memory, the ability to remember that word on the tip of your tongue. Simply stretching had no effect on any of these things.

Further studies showed that a combination of strength and aerobic exercise led to more improvement in executive function than strength training alone. The benefits were later seen in healthy adults, too. Combining exercise types might be particularly powerful because strength training triggers the release of a molecule called insulin-like growth factor-1. This hormone is known to affect communication between brain cells, which is the basis of how we learn, and promotes the growth of new neurons and blood vessels. Meanwhile, aerobic exercise boosts a protein called brain-derived neurotrophic factor, which also triggers the growth of neurons and helps them resist age-related decline. Essentially, a combination of both exercises gives you a more potent cocktail of beneficial brain chemicals.

There's something in it for the kids, too. If you want them to focus for an hour, on a maths test, say, the best bet is to let them have a quick runaround first. That's according to studies that show a twenty-minute walk, sprint or skip has immediate effects on children's attention, executive function and achievement in maths and reading tests. Don't put too much pressure on what they're doing though: exercise that is highly structured or focused on a specific skill, such as a tennis serve, hampers later attention.

That's not to say there isn't a place for these more specific skills in a junior brain-training exercise regime, because they do seem to build up attention span more gradually over the long-term. Twice-weekly sessions of gymnastics or basketball practice, for instance, help children do better in tests that require concentration and the ability to ignore distractions. It might have something to do with the cerebellum, the wrinkly cauliflower-like structure at the base of the brain, which is activated when performing complicated movements, but also works together with the frontal lobe, an area known as the brain's 'control panel', to improve attention. Doing one therefore improves the other.

As an adult you might want to consider adding in another kind of childhood activity to your weekly workout, such as climbing trees or running barefoot. In fact, any exercise that challenges your sense of proprioception (the position and orientation of your body) and some other element, such as navigation, calculation or locomotion – basically, where you need to balance

and think all at the same time – is particularly beneficial. A
good example is surfing, where you have to focus attention on
staying on the board at the same time as judging the best position to catch a wave and determine whether another surfer
might be in your way.

These kinds of exercise have a dramatic effect on our working
memory. This is the ability to hold on to information and
manipulate it in our minds at the same time, and allows us to
prioritise and process information and ignore what is irrelevant.
It influences nearly everything we do, from the classroom to
the boardroom. It can even boost creativity – there's a section
in Chapter 11 if you're particularly interested in that.

And while you might not get an immediate brain-boost from
doing stretching and toning exercises like yoga, there are
numerous long-term benefits. A growing number of studies
show that yoga and mindful meditation – something I discussed
in Chapter 2 – can help produce increased feelings of calmness,
and over time can help with anxiety and depression. One study
of yogis who had been practising for many years found that
some brain regions were remarkably well preserved compared
with those of healthy controls that were the same age, gender,
race and education level. The researchers even commented that
the fifty-year-old yogi brain looked more like a twenty-five-
year-old's.

If you're still unsure which type of exercise to pick, there's
some overlap between the different exercises and benefits, so
take this simple piece of advice: choose the one you enjoy the
most. The best exercise is the kind you'll actually do.

HOW TO TRAIN YOUR BRAIN TO EAT HEALTHIER

We all know that exercise isn't going to take you far if you spend the rest of your day pigging out on fatty foods and takeaways. I'll hold my hands up – in the past nine months I have eaten more chocolate than I care to admit. I could of course blame it on being pregnant, but I've actually got another excuse – as do you. When it comes to food, as much as we'd like to think we direct our behaviours we're not as in control as you might think. We've already seen in Chapter 2 how certain kinds of food can affect our mood. Now, however we're going to concentrate on how to overcome those strange forces of the mind that can lead our diets astray, even when we know what's best.

To unravel this mystery, first it's useful to understand our basic relationship with food. At its simplest, signals between the gut and brain tell us when we're hungry, and when we are full. But the drive to eat gets more complicated than this. Some of our culinary desires are down to that feeling of pleasure, mediated by the brain's reward centres, that we get from eating calorie-dense food like a chocolate bar or a fresh, jammy doughnut. And our gut, and the microbes inside, also manipulate what we crave and determine the way we see food.

So it's no good just telling you to eat better: that's like telling an asthmatic to breathe better. Instead, we have to look a little deeper at those drives. The hunger system is mediated by hormones from the gut and from fat cells, which send information to the brain about when we last ate and how hungry we should feel. We might eat little one day, and binge the next, but this system tends to ensure our body weight stays stable.

Our hormonal reward system is very concerned with what types of food we eat. At its heart is the dopamine pathway, which seems to respond most strongly to foods that are high in fat and sugar. This is natural and necessary: it evolved to prompt us to seek out such big-bang-for-their-buck, energy-dense foods, helping us to survive. In evolutionary terms, if we see a high-energy food, it pays to want it right now, because a famine may be round the corner. The problem is that in our modern environment, where food is often abundant and cheap, our reward system works against us, pushing us towards eating sweet and fatty foods even though we already have plentiful energy stores.

The brain also has its own calorie counter that drives our choices without us knowing. In one study, people were shown pictures of fifty foods and asked how many calories they thought each contained, and then invited to bid in an auction for a chance to eat the foods. Regardless of their calorie estimations, which were often inaccurate, the individuals were more likely to bid for the foods that were genuinely the most calorific. MRI scans showed that activity in reward regions of the brain correlated with the true calorific content of foods – the more calories, the greater the reward.[1]

But what if we could trick the brain into thinking it's seeing a calorie-dense treat that we've learned to crave, when it's really a much healthier choice? Well, it seems we can. When you give a group of people the same milkshake, but tell half that it's healthy, while the other half believe they're having an indulgent treat, you can significantly affect their levels of the 'hunger hormone' ghrelin. This normally drops after a meal, but those who thought they had drunk a low-calorie shake showed markedly higher levels of ghrelin afterwards, which left them feeling less full.

Ghrelin doesn't just affect appetite. By signalling food deprivation, the hormone also slows down metabolism, tipping the
body towards storing fat rather than burning it. It makes evolutionary sense to reduce energy consumption when resources are
scarce, but it's bad news when you are trying to lose weight.
When you think you are eating healthily, it's associated with a
sense of deprivation, which may shape your physiological
response. Instead, dieters should cultivate a mindset of indulgence,
savouring the textures and flavours of whatever they are eating.

Researchers have also designed food with the flavour and
appearance of high-calorie treats that are easily digested, but in
fact they were lower-calorie, slowly digested versions – for
instance, a low-calorie pizza made with added fibre. In a small
trial, a group of overweight people went on a six-month diet
based on these foods. Before and after brain scans showed an
increase in activity of reward pathways when the participants
looked at pictures of healthy, low-calorie foods, compared with
a similar group not on the diet. The diet effectively retrained
their brain. Before, they would think of pizza and crave it because
their brain anticipated a rush of calories. Once they started eating
pizza that didn't give them that rush of calories, over time the
reward circuitry adapted so it no longer expects that boost of
carbohydrates. The craving disappears. Although it's not been
tested, it may be possible to recreate this effect by making your
own low-calorie, high-fibre foods at home: vegetable muffins
that look like the sugary versions we're more used to, for example.

Our desire for food is also affected by the little bugs inside
our stomach, our gut microbiome. We've already seen how they
can affect our happiness in Chapter 2. But our gut microbiota
don't just flourish on certain diets, they may also control our
food craving and preferences to serve their own purposes. We

know, for example, that animals' gut flora affects their taste receptors, which changes what foods they prefer to eat. And many of our own gut microbes can produce proteins that mimic gut hormones. Our knowledge of this complex relationship between our gut and brain is in its infancy, but one day we might be able to give someone specific types of bacteria to make them prefer a certain type of healthier food. Until then, the best advice is what I've already given: make sure you have a varied diet so that you make it harder for any one type of bacteria to flourish and exert control.

In the end, we come full circle and return to exercise. As well as helping you improve your health and lose weight if you wish, exercise can also stem cravings. High intensity, interval training appears to work best in this respect. It lowers levels of ghrelin and triggers less hunger than working out for the same amount of time with a different intensity.

Understanding the complex relationship between our body and our mind is difficult. But the good news is that it reveals that expecting people to rely purely on willpower alone is misguided. Reframing your thoughts, eating a wider variety of food and exercising smarter, might be the quicker – and healthier – way to go.

HOW TO HARNESS THE POWER OF PLACEBO

What we've just discussed points us to a deeper truth. It might sound like New Age nonsense, but it's not just that a healthy body is a fast track to a healthy mind. We now have solid evidence that the reverse is also true: with the right mindset you can think yourself faster, fitter, slimmer – and younger.

It's all to do with the placebo effect. Placebos are inert treatments used in clinical trials to test how effective a drug is. You divide your volunteers randomly into two identical groups, giving half a real drug and the other a sugar pill. With no active ingredients the placebo shouldn't have any effect. Yet many studies have shown that placebos can often bring about significant changes, triggering the release of natural painkillers and lowering blood pressure, for example – all because of people's expectations. The mind is so mysterious that these effects can occur even when people are told they are taking a placebo. Not only that, the placebo effect also has an evil twin, the nocebo effect, whereby people can experience side effects such as nausea, rashes and emotional changes from inert treatments.

The power of the placebo extends to our behaviours. For instance, in one study, golfers who thought they were using a professional's putter perceived the hole to be larger and easier to putt, and were more accurate as a result. Likewise, people tricked into thinking they're wearing designer sunglasses can more easily decipher small writing through the glare of bright light than those who think they are wearing less prestigious brands. Even the effects of caffeine might be in part due to your expectations: pure water can increase alertness and raise blood pressure in people who are told it contains caffeine. The upshot is that two people can have identical genes and lifestyles, but one can end up healthier than the other – thanks solely to their thoughts.

Despite knowing about the placebo response for decades, relatively little has been done to harness it to improve our health and well-being. Interested in how this might be done in practice, researchers examined the fitness of eighty-four hotel cleaners in the United States. They suspected that few of the cleaners would be aware of the sheer amount of exercise their job entails, and that this might prevent them from gaining the full benefits of what is in effect an intense physical workout. To manipulate their mindsets, half were given detailed information about the physical demands of their work – such as the fact that hoovering burns 200 calories an hour – and were informed that their activity met the US surgeon general's exercise recommendations.

One month later, despite reporting no change to their diet or activity outside work, the cleaners who received the information had lost about a kilogram each, and their average blood pressure had dropped from elevated to normal. The other group showed no difference. It was, admittedly, a small study and the team didn't record actual behaviour – it could be that, mindful of potential beneficial effects, the first group of cleaners were putting more oomph into making the beds.

But a follow-up study bolstered the idea that people's expectations directly influence their body's response to exercise. This study used data from health surveys monitoring more than 60,000 people for up to twenty-one years. The researchers found that the 'perceived fitness' of the participants – how they felt compared with the average person – was a better predictor of their risk of mortality than the amount of time they said they spent exercising. Crucially, some of them wore accelerometers for part of the survey period, yet even after taking their actual physical activity into account the influence of their perceived fitness remained. Incredibly, people who took a more pessimistic view of their fitness were up to 71 per cent more likely to die during the survey period compared with those who thought they were more active than average – whatever their exercise routine.

How this works is a bit of a mystery, but it may have something to do with our autonomic nervous system. This is the system that controls things that go on unconsciously, such as our blood pressure, sweat response or digestion. We once thought these beyond our control, but now know that the brain can have some influence over them. For instance, some people are able to control their goosebumps at will while experiments have shown the ability of experienced meditators to raise and lower their body temperature just by thinking about it. Although it's not been investigated in detail yet, a poor perception of your fitness might release stress hormones such as cortisol, which can trigger inflammation and could determine how the body responds to exercise.

If you want to use your mind to boost the effect of exercise, the best approach is not to deceive yourself about your fitness, but not to undervalue what you do, either. You should also avoid comparing yourself critically with your peers, particularly if they are exceptionally sporty.

When it comes to specific types of exercise, there are tricks you can learn to help you improve there too. For instance, I wish I had known more about the psychology of 'bonking' when I was training for the London marathon. Bonking is when a person 'hits the wall', or experiences a sudden onset of debilitating fatigue – which I did around mile 18 in the big race. It doesn't matter if you're a seasoned runner or a complete newbie, it can happen to anyone given the right circumstances. Physiologically, we know it happens when we run out of glycogen, a carbohydrate stored in our muscles and liver that provides a readily available source of energy. At this point, the body switches to using fat stores as fuel. We have loads of fat, but we are much less efficient at turning it into energy than carbs, which means runners can't maintain the intensity they were moving at and have to drop to a walk.

But why do some people hit the wall while others don't? Pacing is important, as this might ease the transition between carbohydrate and fat stores, as is what you eat before and during your run. But psychology also plays a part. A study of male and female marathon runners showed that expecting to hit the wall increased your chances of doing so in a race – even when running history and prior experience of hitting the wall was controlled for.[2]

Other studies suggest there are patterns of thinking that you can hone to minimise your chances of hitting the wall. Runners who daydream during a race are more likely to bonk. And those who spend too much time thinking about breathing or internal discomfort hit the wall sooner.[3] Your best bet is to find some middle ground, focusing on something external like the scenery and spectators, paying attention to race conditions, distance markers and drink stations. Intermittent check-ins about your breathing and how you are feeling are wise, but more attention should be focused externally on the task at hand.

If marathon running seems like a bit too much effort, here's something for the real couch potatoes. A small, but remarkable study by Brian Clark at Ohio University showed that you can use your mind to improve your fitness without lifting a finger. His team used a surgical cast to immobilise the hand and wrist of twenty-nine volunteers for four weeks. For ten minutes a day, half the group sat still while imagining they were performing exercises with their immobilised hand. When the casts were removed, both groups had lost muscle strength in their wrists, but the group who had performed imaginary exercises lost 50 per cent less than the control group. It seems that mental work-outs strengthen pathways in the brain that control muscle movements, which later translates into greater command over the target muscles, increasing their strength.

While we're thinking about thinking, we might also consider how our mindset could be affecting other aspects of our health. Diet is one that I'll come to in the next section – stress is another. I've already discussed stress and how it has both negative and positive sides in Chapter 1. It turns out that fear of stress, or considering it to be debilitating rather than enhancing, is associated with a greater fluctuation in cortisol, the stress hormone. Teaching people methods of controlling their expectations can reduce the actual response.

The dangers of stressing about stress can explain some aspects of insomnia too. We'll investigate the importance of a good night's sleep, and how to ensure you get one, in detail in the next chapter. But about a quarter of people's perceptions of how well they sleep don't correlate with the sleep they actually get, with significant repercussions. People who believe they are insomniacs, even though monitoring their night-time brain activity suggests otherwise, are more likely to experience daytime fatigue, high blood pressure, depression and anxiety. People who really do sleep badly, but don't worry about it, are remarkably free of ill effects during the day. It seems that worrying about poor sleep is doing more damage than the poor sleep itself. The answer, shown in several studies, is to prime yourself or your insomniac partner to think they have slept really well.

Easier said than done, perhaps. But here's one more reason that carving out a more positive attitude is worth the effort: negative beliefs could knock decades off your life. The first clues emerged in 1992, when a group of pensioners were sent to a monastery in New Hampshire and told to act as if they were twenty-two years younger for the duration of their stay. The retreat was decorated as if the year were 1959, and filled with music, films and memorabilia from the era. It contained no mirrors, only pictures of their younger selves. After just five days, the pensioners' arthritis had improved, their posture was more upright and their thinking – as measured by an IQ test – was sharper.

Inspired by this study, other teams have since shown that our attitudes really can influence how our bodies fare over time. Overall, people who view ageing positively live 7.5 years longer than those who associate it with frailty and senility. Negative perceptions of ageing are not merely the result of poor health; they can foreshadow symptoms by as much as thirty-eight years.

Admittedly, people with a pessimistic view of ageing are less active and less likely to seek healthcare when they need it. However, many studies suggest this isn't the full explanation. It seems that people with rosier beliefs about ageing react less to stress and are less likely to develop inflammation – both of which would mean that they age more slowly. They are also less likely to develop brain changes associated with Alzheimer's disease than those who view ageing negatively.

The good news is that attitudes can be changed at any age. In one study, participants aged between sixty-one and ninety-nine played a computer game while positive age-related words such as 'wise', 'mature' and 'experienced' flashed briefly on the screen. Although they did not consciously register the words, their perceptions of ageing had significantly improved after four sessions, as had their physical well-being.

Of course, a positive mindset is not a panacea, nor is it always easy. But if you can increase the benefit you're already getting through exercise, a balanced diet, relaxation and a good night's sleep, merely by thinking positively, surely it's worth the effort?

TOP TIPS FOR LIVING HEALTHIER, LONGER

🔧 Aim for 150 minutes of aerobic exercise (running, brisk walking, dancing) each week, but don't forget to add two bouts of strength or resistance exercise (lifting children, pumping iron, digging in the garden). Both kinds of fitness help improve different aspects of your physical and mental health and boost the effect of either alone.

🔧 Different exercises can improve different cognitive functions. Lift weights to help you problem-solve, run to improve your attention and memory, and do yoga to lower your anxiety and risk of depression.

🔧 A positive mindset can increase the benefits you get from exercise and add several years to your lifespan.

🔧 Learn to curb your cravings with short bouts of high-intensity exercise and eat a varied diet to keep your gut bacteria from exerting an undue influence over your body and brain.

🔧 Try cutting down on your calories a few days each week. Calorie restriction is looking like a promising way to avoid age-related diseases for longer.

7

HOW TO
SLEEP WELL

I REMEMBER THIS particular night so well. It was 1 September 2018, and my daughter was exactly one month old. I had just laid my head down to sleep when she woke up, screaming to be fed. It wasn't an unusual occurrence, but I'd had a particularly gruelling week and I'd reached peak exhaustion. After a forty-minute feed, I laid her back down. I fell asleep instantly. Just under an hour later and I was woken by her crying out for more food. Bleary-eyed and still sore from my C-section I rolled over and fed her again. Twenty minutes later and I was back asleep. An hour later, though, the same thing happened. And the next hour, and the next.

Around 4 a.m. I woke up, startled. Lying in the middle of the bed was my daughter. I'd fallen asleep with her next to me. I tried to pick her up and my husband gave a startle. It was actually his head and his arm that I was trying to manoeuvre back into the cot, where my baby was actually safe and sound, sleeping. It was the first of hundreds of similar hallucinations of having left her in the bed that I was to have over the next six months.

When morning came, I burst into tears and told my husband that I couldn't do this any more. The tiredness had enveloped me in a black cloud of misery. I felt physically sick, I was overwhelmingly sad, I couldn't think straight. All because of a lack of sleep.

If you've ever been through a period of sleep deprivation, whether because of children, night shifts, stress or insomnia, I'm sure you can relate to this. Sleep is a gift that can be so pleasurable when we get enough of it, but it can cause such havoc to our physical and mental health when we don't. And for something that we do so much of, it's still such a complex and mysterious thing.

I've always wondered, for instance, why my sleep patterns are so different from my friend Emily's. If my children would allow it, I would sleep until 10 a.m. Emily would recoil in horror at this statement. She happily wakes at 6 a.m., bright as a lark, making fresh bread, hours before I've even stirred. Am I just lazier than her? Should I be making more use of the dawn chorus? Then there's my husband, who often claims not to have slept at all, but I know for certain that he's been fast asleep for hours, because he's been putting me off my own sleep with his gentle snores.

We have a bit of an obsession about sleep, whether we're worrying about missing too much, or one of those proud to need so little. But we're right to be obsessed. Sleep is as vital for life as food or water. Along with diet and exercise, it has been labelled the third pillar of good health. Even that is under-selling it: sleep is the foundation on which these two other pillars rest.

Lab rats deprived of sleep die within a month, while people who inherit the rare disease 'fatal familial insomnia' meet the same fate over time. A total lack of sleep kills a person quicker than starvation, and in milder forms can turn all of us into walking zombies, cognitively and physically. There is no tissue in the body or process in the brain that is not significantly enhanced by sleep, or impaired when we don't get enough. Lack of sleep is associated with problems with our immune system and appetite, with obesity and diabetes, and with mental health problems like depression, schizophrenia and Alzheimer's disease.

If a lack of sleep is keeping you up at night, making you hallucinate or affecting your health, read on to find out how to make the most of your sleep, how to avoid insomnia, how much rest you really need and how to catch up on lost zeds.

HOW TO UNDERSTAND SLEEP

All animals sleep. Even cultured brain cells in a Petri dish spontaneously enter a sleep-like pattern. The average person spends a third of their life asleep. It's clearly important. But what is it for?

'The only known function of sleep is to cure sleepiness,' Harvard sleep scientist Allan Hobson once joked. We know a bit more than that of course, but our interest in sleep as a biological concept only really began in the 1950s, when PhD student Eugene Aserinsky stared at a sleeping patient and noticed their frantic, jerky eye movements. He'd been expecting far smaller movements, given that, back then, sleep was thought to be a passive, uninteresting state of unconsciousness. His discovery kick-started the area of sleep science – a surprisingly difficult and misunderstood discipline, considering that sleep has such a profound effect on our physical and mental health.

Aserinsky's observations were of what we now call REM, or rapid eye movement, sleep. This is when we experience our most emotionally charged and vivid dreams. It is one of numerous phases that we cycle through when asleep. The others are lumped together as non-REM sleep. This is thought to be a deeper type of sleep, where our dreams are less clear and less memorable.

Our non-REM sleep can be split into various stages. Stage 1 occurs just after we nod off, and lasts only a few minutes. If someone shakes you awake during this time, you'll probably think you weren't even asleep. But it can also feature some pretty strange hallucinations and even something called 'exploding head syndrome', where you hear a loud gunshot or bomb noise that startles you awake. The cause remains elusive, but may be something to do with parts of your brain entering a dream-like state, while others remain conscious. Stage 2 is pretty uneventful. Our brainwaves, which are the coordinated electrical activity resulting from several brain cells firing at once, slow down and are interrupted only by brief bursts of electrical activity, and there is no dreaming.

About twenty-five minutes later we transition into Stage 3, known as slow-wave sleep. This is the deepest and most restorative stage of sleep, in which our heartbeat slows and our brainwaves become long and regular. Only minimal dreaming, if any, occurs during this time. If we wake during this cycle it can take up to an hour to shake off that horrible grogginess you feel. This stage lasts up to forty minutes, before you cycle back through Stage 2 and into REM.

This is when your dreams start getting interesting. Your muscles are paralysed now, which is probably to stop you acting out your dreams. When you scan someone's brain during this stage of sleep it can look very similar to when they are awake. This is thought to be the time when the brain processes memories and emotions. Each round of REM sleep is followed by Stage 1 and then you go through the phases all over again. With each phase, your REM sleep lasts longer, until you've had around two hours of it altogether. Why these cycles exist and exactly what's happening in your brain during them is a matter of great debate. Sleep is clearly restorative, and the best way to understand what it might be doing for us is to look at what happens when we get too little.

A lot, is the short answer. As I've mentioned, poor sleep is a risk factor for pretty much every disease going, from cancer to Alzheimer's, but it also interferes with your ability to fight infection, depletes your willpower and stops you being able to pay attention or make a decent plan. If anyone debates the importance of sleep, tell them that driving on less than five hours sleep leaves you with the same cognitive impairment as having a blood alcohol content of 0.1 per cent – more than the drink-drive limit in a lot of countries. If they still need convincing, tell them that chronic sleeplessness of four hours for six nights in a row increases blood pressure and insulin resistance, and makes you produce half the normal number of antibodies in response to a virus.

Despite – or in fact because of – all of this, for most of us, lack of sleep is not deadly. The longer you stay awake, the greater the build-up of a chemical called adenosine in the brain, which sends signals that increase your desire for sleep. After about sixteen hours, that desire becomes overwhelming. Unless you've drunk a lot of caffeine, which keeps your awake by blocking adenosine receptors in the brain, you'll fall straight to sleep.

But why do we need to sleep at all? There are several ideas, of course, ranging from obvious ones about restoration and recovery to more elaborate theories dealing with memory-processing. But if we look at how levels of sleep can influence Alzheimer's, one of its main functions is revealed. Our non-REM sleep gets worse as we age, and it is especially disrupted in this disease. Alzheimer's is associated with a build-up of a protein called beta-amyloid, which forms sticky clump-like plaques in the brain. What exactly these plaques are doing and their precise association with Alzheimer's are still matters of debate – I'll cover a new and surprising theory in Chapter 9 when I deal with how to stave off mental decline. But their presence seems to stop brain cells working properly and eventually kills them. The plaques are particularly prevalent in the middle of the frontal lobe, an essential region for generating deep sleep.

This is a two-way street, though. While we sleep, our brain's glymphatic system – a type of waste disposal system for all the harmful products of brain activity that have built up through the day – kicks into high gear. It expels twenty times more debris while we are unconscious than it does while we're awake, and this debris includes amyloid proteins. And so the vicious circle continues: lack of sleep causes amyloid to build up in the brain, which inhibits the exact kind of deep sleep you need to remove the amyloid. More amyloid, less deep sleep, and so on.

A lack of sleep will increase your risk of Alzheimer's in the long term. The good news, however, is that clinical trials suggest that fixing sleep problems can slow the rate of cognitive decline and delay the onset of Alzheimer's by up to ten years.

But what of REM sleep – what's that all about? Some researchers have proposed that this is the brain booting up to test out the repairs it made during non-REM sleep. Others focus on its role in processing memories and emotions. What's clear is that it's not there by accident. When you deprive mice of REM sleep, they don't seem to be able to consolidate memories about tasks they have learned the day before.

REM sleep also seems to have a role in creativity, something I'll cover in more detail in Chapter 11. When you measure people's creativity after allowing them to rest quietly, to have a nap of non-REM sleep or experience REM sleep, those in the last group showed greater improvement in creative tests than the others. Although it's not clear exactly what the mechanisms are behind this phenomenon, it makes sense when you consider that REM sleep seems to put the brain into a state where it makes lots of unusual associations in our weird and wonderful dreams – unusual associations being a pillar of creativity.

Dreaming itself might allow the brain to monitor your emotional response to decisions or memories and help you to decide which are useful or not. Studies show that decisions are indeed easier to make if you have slept on them. Some researchers go as far as to say that your REM sleep processes all your undigested psychological material from the day. They believe it is a kind of 'overnight therapy' that strips out the emotion from traumatic or potentially anxiety-inducing memories. Some small studies back this up, by showing that people who sleep and dream well heal more quickly from emotional hardship.

Hopefully by now you're convinced that you should prioritise your sleep, for your physical and mental health. But how much do we need and how do we improve it?

HOW TO KNOW HOW MUCH SLEEP YOU NEED

In 2015, I spoke to Abby Ross, a retired psychologist from Miami, Florida, whose claim to fame is that she sleeps for only four hours a night. She told me it felt like she lives two lives. While the rest of the world was sleeping, she was completing university in half the time, getting on with her work in peace and quiet, and fitting in extra exercise that resulted in her completing thirty-seven marathons and several ultra-marathons.

If you believe the stories, UK Prime Minister Margaret Thatcher also survived on next to no sleep. When all these hours are added together, 'short-sleepers' can have sixty days of extra time to themselves every year. That might sound bonkers to you. And indeed, short-sleepers are certainly in the minority – probably less than 3 per cent of people can manage on this amount of time.

The rest of us know that eight hours is the magic number for a decent night's sleep. Or is it? Nobody really seems to know where this number came from. In questionnaires, adults tend to say they sleep for between seven and nine hours a night, which might explain why eight has become a rule of thumb. Teenagers sleep between eight and ten hours, and children a lot more. Babies can sleep up to seventeen hours a day.

In fact, the eight-hour rule has no basis in our evolutionary past. Tribal cultures with no access to electricity tend to get around six to seven hours' sleep, without any health implications. That said, seven hours does seem to be a safe minimum require-ment in other parts of the world. An analysis in the United States concluded that regularly getting less sleep than that increases the risk of obesity, heart disease, depression and early death, and recommended that all adults aim for at least this.

By this benchmark, a lot of us are walking around in a state of sleep deprivation. The US Centers for Disease Control and Prevention estimates that 35 per cent of American adults are getting less than seven hours a night, and a survey in the UK found that the average was 6.8 hours.

You might think you don't need as much sleep in the normal course of things because at the weekend, on holiday or during a break from the kids, you sleep longer. We think we're catching up on 'lost' sleep but it's not necessarily true. In the same way that we eat more purely for indulgence's sake, the same could apply to sleep too. Why not go all out when you can: it can be very pleasurable to fall back to sleep when you would otherwise have made an effort to get out of bed. It doesn't mean you needed that extra hour or so.

The amount of sleep we actually need almost certainly comes down to our genes. A recent study of more than 50,000 people found one gene variant that added 3.1 minutes of sleep needed for every copy you have. Imagine what a whole host of genes working together has on your need for snoozing. Likewise, people who need very little sleep often have mutations in a gene called DEC2. When mice were bred to express this mutation, they also slept less but performed just as well as regular mice in physical and mental exercises. What this gene does, and how to replicate its action in the rest of us, is unfortunately still unknown.

Taking these individual genetic variations into account, the US National Sleep Foundation updated their guidelines, and came up with a recommended range of seven to nine hours for adults, but with added leeway of an hour either side.

So how much is enough for you? A general rule of thumb is that you shouldn't need an alarm clock to wake you up in the morning. It's something I've recently discovered for myself. I no longer set an alarm in the knowledge that my daughter will never sleep past 7.30 a.m., yet almost every morning I wake around 7 a.m.

While I'd personally love to go back to the ten-hour nights of my university days, be warned: you can have too much of a good thing. There does seem to be a sweet spot for maximising sleep for health. While getting too little can make you ill, regularly getting more than eight hours could also increase your risk of early death. Why this is remains a mystery, but could be down to the simple fact that when we are asleep we are moving very little. As we saw in Chapter 6, inactivity is a rapid route to bad health. And although this might not matter if you are active during the day, it could be that people who spend more time asleep do less exercise, possibly because they simply have less time.

You can test how long it takes you to get to sleep. On average this should be around ten to fifteen minutes. If it is quicker, then you might be sleep deprived. Of course, you might not need as much sleep as you think. Many of us sleep for a long time out of habit or boredom, so it might be worth trying to cut down and see how you feel.

HOW TO IMPROVE THE QUALITY OF YOUR SLEEP

Nothing beats a bath and what we in the Thomson household like to call 'sleepy tea' as a sure-fire way of sending me quickly off to the land of nod. The tea is made from a herb called valerian root, which has many reputed sleep-inducing effects. Unfortunately, systematic reviews of the studies that investigate it as a sleep aid have concluded that while safe, it probably has no more effect on sleep than a placebo. But while my tea-induced sleep is more psychological than physiological, there are other more scientifically sound things you can do to help improve your ability to slumber.

First, it might seem obvious, but make sure you are comfort-able – not just with a decent mattress, but feeling safe and happy in your place of sleep. You might have noticed that you don't sleep too well when you are in a new environment: the first day of a holiday, for instance. When people had their brains scanned while sleeping in unfamiliar places, researchers found that parts of one side of the brain remained active while the participants were asleep that weren't active as that place became more familiar. It might be an evolutionary adaptation, a kind of night watch, keeping part of your brain alert to make sure your new environment is safe.

Next, keep it cool. If you're anything like me and my husband you have a constant battle over the thermostat during the day. But when it comes to the temperature of your night-time environment, there is a sweet spot of around 18.5 degrees Celsius at which most people can sleep optimally. That's lower than the sort of room temperature at which we function best when active, and helps our body's core temperature decrease by about 1.2 degrees, which it needs to do to get to sleep.

It sounds counter-intuitive, but this might be why my bedtime bath helps me sleep. It's not your periphery but your core temperature that needs to drop, and warming your skin in a bath or by wearing a specially designed 'sleep suit' actually helps the body to release heat. People who have sleep disorders who warm their skin in this way wake up less during the night and have more restorative deep sleep.

Next, keep it dim. You might be one of the vast number of people who snuggle up with their phone as soon as they slip under the covers, but phones are truly the worst bedfellow. Tablets, phones and laptops generate lots of short-wavelength blue light, which interferes with our production of the sleep hormone melatonin. We normally start to produce this in the mid to late evening, but one study showed that using your screen for two hours before bed reduced melatonin concentration by 22 per cent.

Screen time just before bed can also mess with your whole sleep cycle. Our bodies march to the beat of internal timekeepers called circadian rhythms, which control cycles of physical and mental changes that are thought to prepare our brain and body for the tasks we're likely to encounter at certain times of the day. Our sleep–wake cycle is one such rhythm. Staring at a lot of blue light before bed can push it forward, meaning it's harder to get to sleep. Once we are asleep, we have less time to get through all of our normal cycles of sleep. Because our REM phases increase as the night goes on, cutting our sleep short might mean we get less benefit from this type of sleep.

Don't worry about television before bed – we sit far enough away from it to mitigate any problem. Also, if you're not using your digital device for more than two hours, you're probably not doing too much damage: less than this doesn't seem to significantly reduce melatonin production. If you want to cut your exposure down, you could try using an app that strips out much of the sleep-robbing light, or switches it to a dimmer, red light that doesn't cause the same issues.

Keeping the environment dark is a good idea. People who live in cities with lots of light pollution tend to go to sleep and wake later than people who sleep in more natural environments. They also feel groggier in the morning, and are less satisfied with their sleep quality. This might be because there's more chance of noise waking them during the night or something else entirely, such as having a more stressful career perhaps, but a range of studies suggests light is certainly involved. Get yourself a blackout curtain or eye mask and sleep easier.

Whatever you do to cure your insomnia, stay away from alcohol. It might be tempting to have a tipple before bed and it's true it is a sedative of sorts. But sedation is not sleep. Drinking before bed seems to disrupt deep, non-REM sleep and causes more wakefulness in the second half of the night.

Sleepy tea and alcohol may be out, but there are other drinks that might be more conducive to a good night's rest. Sour cherry has been shown to be rich in melatonin, and in one study adults who drank it twice a day for a week slept for an average of thirty-four minutes extra each night and napped less during the day. Other foods might work in a similar way, says the US National Sleep Foundation, such as almonds, walnuts and pineapple.

You could also try eating foods that are rich in the amino acid tryptophan, such as nuts or cheese. Tryptophan helps in the production of serotonin, a happiness chemical, and serotonin is a necessary building block the body uses to create melatonin. Low serotonin levels are thought to contribute to insomnia and several studies have shown that eating foods rich in tryptophan makes people feel sleepier.

There are plenty of herbal sleeping pills, ginger tea and essential oils like lavender and hops that are touted as sleep aids. As with valerian, however, the evidence is mixed and studies are often weakly controlled, meaning it's hard to give any concrete advice on their worth.

Finally, one of the most effective ways to improve your sleep is to keep your waking time consistent. Your brain prepares to wake up around ninety minutes before actually waking up, so if you chop and change your brain will find it harder to use the time asleep as efficiently as it could. The same goes for pressing the snooze button multiple times: you may be messing with your finely tuned sleep architecture.

One last thought. Ignore anyone who makes you feel bad for needing more sleep than they do. We've touched upon the fact that scientists believe that the amount of sleep we need is as genetically determined as our shoe size. Some of us need six hours, while others need ten. We're pretty bad at working out how much sleep we've actually had, often underestimating it. Stressing about not sleeping, or having an unrealistic expectation of how many hours we should be getting, could be doing more harm than good. Finding out what your ideal sleep time is and trying to work your life around it could be the best medicine for the perfect night's sleep.

HOW TO NAP LIKE A PRO

No matter how well we might sleep in general, sometimes we burn the candle at both ends and just need a nap. Don't be ashamed: once dismissed as a sign of laziness, it's now clear that taking forty winks is a great way to improve your performance.

But not all naps are created equal. A 'nano-nap', lasting just ten minutes, can boost alertness, concentration and attention for as much as four hours. Take twenty minutes and you increase your powers of memory and recall, too. Either way, you are unlikely to enter the deeper stages of sleep, so will avoid that groggy feeling that can occur when waking from deep sleep. On the flip side, you won't get the brain-repairing benefits of deep sleep. But light sleep turns out to be more important than we thought. The secret comes down to the little bursts of electrical activity that occur during light sleep called 'sleep spindles', which seem to have a role in learning and memory.

A slightly longer nap could also improve your equanimity. If you are feeling emotional, try snoozing for forty-five minutes or more. This should take you through a stage of REM sleep. Brain scans of people following a REM sleep nap showed more positive responses to images and to pleasant experiences. If you're looking for the biggest boost to learning, opt for a nap of between sixty and ninety minutes. Research suggests that this aids learning by shifting memories from short-term storage in the brain's hippocampus to the prefrontal cortex – a bit like clearing space from your computer on to a USB memory stick. As well as helping you to retain factual information, longer naps can increase motor memory, which is useful for learning new skills in sport or playing a musical instrument.

Bear in mind, though, the time of day you nap may affect the type of sleep you get. As we've seen, we get more REM sleep as the night goes on. A morning nap is more likely to contain REM sleep as the brain still has a preference for it. By the afternoon you'll get less. Morning naps therefore tend to contain more emotionally calming dream sleep and afternoon naps more restorative and memory-boosting deep sleep. That said, there's not a lot of research on this particular area of napping, and it's still to be proven whether you can really hack your naps in this way. Besides, for all of us, the urge to nap is strongest at one particular time of day: after lunch.

Don't blame what you've eaten – blame your circadian rhythms. Everyone has a pre-programmed drop in alertness around this time, and the European cultures that have maintained a siesta in their way of life may be on to something. Greek men who did away with a post-lunch nap, for example, had worse cardiovascular health and increased rates of cancer. It could be down to a loss of deep sleep, which lowers blood pressure and the speed of the heart so that you wake up with a better managed cardiovascular system.

One last tip: if you want to keep a nap short, drink a cup of coffee immediately before trying to doze off. The effects should kick in around twenty minutes later, wiping away any chance of sleep inertia and leaving you raring to go.

HOW TO LEARN WHILE YOU SLEEP

The idea that we could learn while asleep was once the stuff of dystopian science fiction. In both Aldous Huxley's *Brave New World* and Anthony Burgess's *A Clockwork Orange*, authoritarian regimes use sleep learning to brainwash the lead characters. For a while it also became the mainstay of numerous teach-yourself courses that claimed you could learn a foreign language while you napped. Experiments in the 1950s to test this idea produced some promising results initially, although critics wondered whether the subjects might be feigning sleep as the recordings played. Sure enough, when researchers started measuring partici-pants' brainwaves to make sure they were truly in the land of nod, the effects all but disappeared: those who truly were asleep seemed to have learned nothing.[1]

You'd be forgiven for thinking this was case closed. Yet today, after more than half a century in disrepute, sleep learning is experiencing something of a revival. Ingenious experiments are revealing that our sleeping brains can absorb new information – under the right circumstances.

For a start, we know that when we sleep the brain doesn't completely shut off. One aspect of the sleeping brain that we have not really touched upon is the way it appears to review and store memories, replaying the experiences of the day to preserve important information. The fact that memory consoli-dation happens while we sleep made some people wonder if we could exert any control over the process. To test the idea, eighteen people were asked to play a memory game shortly before going to bed. Each person learned the locations of fifteen pairs of cards on a computer screen while smelling the scent of roses. Then, while sleeping, they were re-exposed to the scent, which the researchers thought would cue memories associated with what they had learned. Sure enough, people recalled more card pairs after being exposed to the odour than after sleeping without the smell. Another study showed that people who learned to play a simple song on a video game similar to *Guitar Hero* were better able to play that melody upon waking if it had been played quietly as they slept.

In 2011, scientists made another discovery. Volunteers who learned a set of word pairs believing they would be tested the next morning performed better than those who weren't informed of the test, or who were informed but didn't sleep. It suggested that the mere expectation that a memory will be important in future is enough to incite the sleeping brain to replay and strengthen it.

Researchers then wondered whether you could do the opposite, and unlearn a memory while you slept. Sure enough, studies showed that you could do just that — helping people unlearn ingrained prejudices, for instance, such as the idea that women are bad at science, while they slept.

Meanwhile, researchers at Northwestern University in the United States managed to erase a bad memory while volunteers slumbered. They showed their participants images of faces while giving them a mild electric shock and an odour: mint, lemon or pine. Once they had learned to associate some of the faces with pain, the participants slept, while the researchers exposed them to just the smells — no shock this time. At first these triggered anxiety, as measured by microscopic sweat on their skin, but gradually the fear diminished. When they awoke, they were less anxious in response to the images. Volunteers who underwent the same procedure but without sleeping didn't lose their fear.

With the realisation that you could enhance or reduce the

retention of specific memories, people began to question the dogma that the brain can't learn new information as we sleep. Perhaps the scientists in the 1950s were actually on to something.

Two recent studies are even more convincing. In the first, researchers attempted to teach a simple association during sleep – linking a sound with an odour. Humans unconsciously inhale deeply for a pleasant odour and sniff weakly for an unpleasant one. So sleeping volunteers were exposed to an audio tone paired with a pleasant odour (a deodorant or shampoo) and a second tone paired with an unpleasant odour (rotten fish or meat). Although they had no conscious awareness of either the tones or smells, upon waking the volunteers sniffed deeply when they heard the tone linked with the nice smell and shallowly on hearing the other – without any odours being present. It showed that people could learn an association in their sleep, and that this association could be retrieved while awake.

A second experiment tested the utility of the finding by inviting sixty-six smokers to spend a night in a sleep lab, during which they were exposed to the smell of cigarettes paired with the smell of either rotten fish or rotten eggs. Lo and behold, they smoked 30 per cent less the week after the experiment than the week before.[2] The same procedure conducted while participants were awake didn't result in altered behaviour. It suggests that there is something special about how our brains process memories during sleep that enhances learned associations, though we don't yet know what it is.

What we do know is that learning rarely occurs during REM sleep and mostly in non-REM sleep. One possible explanation is that the slow oscillations of brain cells cement memories. It could also be a protective mechanism so that memories are not formed while we are dreaming vividly.

Whether we can learn more than simple associations is still a work in progress. We know that the part of the brain responsible for processing audible information is active during sleep and responds preferentially to more meaningful information. For instance, you're more likely to wake up if someone calls your name or yells 'Fire!' than if they call someone else's name or a nonsensical shout.

In one final and tantalising experiment, volunteers who were hooked up to machines that scanned their brainwaves were asked to classify spoken words as either animals or objects by pressing a button with either hand. After beginning the experiment, participants reclined in a darkened room and drifted off to sleep as they continued to classify words. At some point, they stopped pressing the buttons, but their brains didn't stop categorising. The hemispheres associated with pressing a button continued to light up correctly in response to the spoken words. Their unconscious brains were still absorbing and processing meaningful information, though much more slowly than when awake.

It demonstrates that during sleep you can extract meaning from acoustic information, and also prepare a response or make a decision. It implies that if we could work out how to deliver the information, the sleeping brain may be ready and able to learn it. It's early days and we need to proceed with caution. As we know, good sleep is a requirement, not an option, and there may be a trade-off. If we try to induce learning while the brain is supposed to be resting, we would undoubtedly reduce the benefits we get from sleeping.

Nevertheless, keep an eye on this area of research. Some day soon, we could be using sleep to unlearn deep-rooted prejudices, alter bad habits or learn new positive associations with certain foods or experiences. Perhaps one day we'll even use sleep to help us learn a new language. But for now, best stick to getting your beauty sleep without any distractions.

TOP TIPS FOR SLEEPING WELL

🔧 Get up at the same time every day so that your brain can start to prepare you for waking at the most refreshing moment.

🔧 Avoid prolonged periods staring at a tablet, phone or laptop in the hours before bed – the blue light you are exposed to can diminish your melatonin levels, a vital sleep hormone.

🔧 If you suffer from insomnia, drink sour cherry juice twice a day: it's rich in melatonin and will help you sleep.

🔧 Don't compare yourself to others, everyone needs different amount of sleep. This is down to genetics not laziness.

🔧 A short nano-nap of ten minutes can boost alertness, concentration and attention for as much as four hours.

8

HOW TO MAKE AND BREAK HABITS

I BITE MY nails. It's not the worst habit in the world, but just once I'd like to get a manicure without having to apologise for the state of the short, jagged monstrosities at the end of my digits. I think it started towards the end of primary school. Up to that point I was probably too busy sucking my thumb to pay much attention to any other part of my hand.

I've tried to stop, of course, just like I managed to stop sucking my thumb when it became too embarrassing, but it's been a lot more difficult. With my thumb-sucking, I went cold turkey after coming to the realisation that I was far too old for the habit. I had a lot of motivation: I was already teased in the playground for having huge glasses, and for being a bit of a swot. I didn't need to give the little bullies any more ammunition.

But I don't have such motivation for my nail-biting. Sometimes, before a big event, like my wedding, I've managed to go a whole month without chewing. But without that kind of mental reward, I'm back to biting. Sure, there is practical stuff I could do to help: nasty-tasting nail varnish or fake nails would put an end to it. But both of those are a bit of a faff. In all honesty, I've not been particularly driven to seek out a solution.

Recently, though, I've been thinking more about my bad habits and why I've got into certain patterns of behaviour. Because it's not just the nails. I've also got into the habit of missing my weekly yoga class, and of watching Netflix far too late into the night. And in the past three minutes, I've probably reached for my phone half a dozen times, for no particular reason. Then again, recently I've also acquired the good habit of visiting my allotment every other day, which not only adds to my weekly exercise, tops up my exposure to light and greenery, but is providing my family with loads of lovely vegetables and flowers.

But the problem with habits is, whether they are good or bad, they are often unconscious. Much of what we do in our day-to-day lives – walking to work, checking our phone, eating breakfast – is habitual, around 40 per cent, in fact. It's only when habits start to have an overtly positive or negative impact on our lives that we tend to give them any thought. Of course if a habit turns into something we have no control over, to the point where it could be harmful, it becomes known as an addiction.

Having given more thought to my own habits, I wanted to find out how I could do more to enhance the good ones and get rid of the bad. The world of self-help thrives on blogs, books and classes that can help you control your habits, particularly things like smoking, overeating or exercising. But a lot of well-trodden advice is based on outdated or anecdotal evidence. Even some of our oldest and most popular methods of quitting smoking have only recently been tested in well-controlled trials, with mixed results. Then there's oft-repeated 'facts' that are churned out as truths – that it takes twenty-one days to form a habit or get rid of an old one, say. Their origins are unknown, and they generally turn out to be complete nonsense when properly studied. (You can find out how long it might actually take you to change your habits later in the chapter.)

The good news is that in the past decade, neuroscientists have learned a lot about what happens in the brain when habits form. Habits might rule our daily lives, but understanding how they become ingrained in the brain, and what we can do to dislodge them, can help you turn the behaviours you want to keep into permanent habits, and drop those that cause you harm for good. Ask yourself the right questions, know how to boost your will-power and make a few small changes, and you'll soon be making or breaking your habits, I promise. Although, admittedly, my nails are still a work in progress.

HOW TO UNDERSTAND YOUR AUTOPILOT

Deep in the brain is a region known as the striatum. It is intimately involved in voluntary movement, as well as having an especially important role in creating a sense of reward for certain behaviours. It integrates our actions with the pleasure we get from performing them, and is involved in reinforcing neural pathways that make us progress from merely enjoying the anticipation of something pleasurable to compulsively seeking it out. It is the brain's habit-forming centre – in some sense, our autopilot.

If we look closer, we see that when we first undertake an action (biting our nails, say, or learning to touch-type or perform a tennis serve), the prefrontal cortex, which is involved in conscious planning of complex tasks and decision-making, communicates with the striatum, which sends the necessary signals to enact the movement. Over time, however, input from the prefrontal circuits fades, and is replaced by loops linking just the striatum to the movement areas of the brain. These loops, together with memory circuits, allow us to carry out the behaviour without having to consciously enact it. That's why practice makes perfect. No thinking is required.

The upside to all this is that we no longer need to focus our attention on a frequent task, freeing up precious processing power to be used for other things. The downside is that once our behaviour becomes a habit, it is less flexible and harder to interrupt: you need conscious intervention to break it again. If it's a good habit, that's a great thing. You don't want it to be easy to forget how to touch-type, play tennis or drive a car. But if you cement a bad habit in the brain, it's equally difficult to get rid of it. You've lost that moment of choice when you can decide not to do something.

Habits can sometimes turn into addictions because they flood the brain with feel-good chemicals such as dopamine, which activate reward pathways. The reward strengthens the neural circuits involved in wanting and performing the habit. It can also stimulate the release of a chemical called glutamate, which is involved in learning and memory, and enhances the pathways between different parts of the brain (such as the striatum) that are responsible for forming cues and rewarding behaviours. All together, this reinforces your desire to do the thing that made you feel so good, even if it's very bad.

All is not lost, however. While habits are etched strongly in our brain's architecture, there is still a little wiggle room. For instance, even when we are performing our most ingrained habits, a small area of the prefrontal cortex is keeping an eye on proceedings, in case we need to take alternative action. You might never need to give much thought to using the brake pedal in your car, but if it stops working, your entire focus of attention easily shifts to the physical act of driving the car. If you're brushing your teeth and suddenly you get pain from your gum, you stop and change your action. No matter how ingrained your habit, your brain remains flexible, and therefore has the potential to rewrite itself – if you know how.

So, how? The next step in changing your habits is to recognise that they don't form in isolation. They are built up in the brain alongside memories and information about the environments in which we tend to most often carry out those habits. In other words, they have triggers.

Think about eating popcorn. Do you ever really fancy it unless you're sat in front of a movie? When researchers gave people week-old popcorn, they ate more of it in a cinema setting, even when they admitted it didn't taste very nice, than they did when given it in a conference room. You can apply this logic to any habit. If you're used to smoking with an alcoholic drink, just the sight of a pub or bar can trigger a craving for a cigarette.

Your brain builds up associations between different cues in our environment very easily – especially when those things make us feel particularly good, or particularly bad. Evolution has given us brains that tend to play it safe in the survival stakes. So, for example, I have an aversion to marshmallows, because I once ate a whole pack and then had acute appendicitis a few hours later. The marshmallows had nothing to do with my appendix erupting, but my brain has formed a strong association between the two that's hard to break. For my brain, it's safer to avoid marshmallows than to risk the same thing happening again.

These associations are happening all the time in the brain: it's the basis for learning and memory. When we smoke and have a glass of beer at the same time, the neurons associated for the reward you get from smoking (we'll get on to that later in the chapter) and those responsible for the concept, taste and smell of a beer, will be actively firing electrical pulses around the brain at the same time. The old saying is that 'neurons that fire together, wire together'. And once two sets of neurons are linked in this way, activating one will make the other more likely to fire: when you see a pub, or taste a beer, you can't help but consider having a cigarette too.

All this goes to say that when you're forming a plan to break a habit, make sure you take account of the kind of cues that might be associated with that habit. Plenty of studies back this up: if you have a better understanding of what context or environment is likely to trigger your craving, it's more likely you'll succeed in stopping it.[1]

This link between our habits and our environment takes us on to a further piece of preparation: there are good times and bad times to break or make a habit, so think about when you're trying to start or stop yours. Studies show that people tend to change habits more easily when their surroundings get disrupted, say when they start a new job or transfer to university. Your old environment is probably strongly linked with your old habits, so changing things starts to weaken those associations you've created in your neural pathways.

Think of it as an overgrown path through the woods to your house. The more you use it the clearer it becomes and the more likely you'll take it every day. If, suddenly, a new path becomes available (you come from a different direction, say), the old path gets used less, becomes overgrown, and we're less likely to take it in the future. A similar thing happens in the brain: if we stop triggering two associations at the same time, the neural pathways between the two become weaker and eventually break down. Even small changes to your routine can help. Little changes to the environment can make a big difference to the brain.

That brings us to a final piece of advice about trying to make or break habits: be kind to yourself. There are sure to be a few slip-ups. The key is not to beat yourself up about them. A few steps backwards shouldn't be a real problem. When a University College London study followed a hundred people as they tried to form new habits, they found no long-term consequences to falling back on old ways. Whatever you do, remember that it will get easier. Every time you practise a new habit or avoid an old one you strengthen and weaken the pathways in the brain that underlie each, making a positive step towards a new you.

It may take time. Exactly how long will differ greatly between individuals. Twenty-one days is certainly optimistic. In fact, there's a huge variation in how long it takes to cement new habits – when tested, the average is about sixty-six days, but one study showed it ranged between 18 and 254.[2] So don't give up, you'll get in the habit eventually – perhaps assisted by some willpower.

HOW TO HARNESS YOUR WILLPOWER

The Cookie Monster in *Sesame Street* isn't known for his self-restraint, but in 2013 the swivel-eyed biscuit fiend experienced a remarkable transformation. Over a series of episodes, he learned to curb his cravings and avoid eating every cookie he saw, in an attempt to gain entry to the Cookie Connoisseurs Club. 'Me want it, but me wait' is how Cookie Monster described his dilemma in a catchy musical number.

Parents of young viewers may have sensed this storyline had a purpose. It was an attempt to tap into the latest research on willpower. Harnessing willpower – our self-control and determination to do something, perhaps against the odds or what we feel to be our immediate self-interest – doesn't just help you to stop your own cravings but has also been linked to many aspects of success in life.

The classic experiment that all willpower research has grown from was the work of psychologist Walter Mischel. In the 1960s, Mischel presented children with a tray of tasty treats, including marshmallows. The kids could either eat one treat immediately, or wait for a few minutes while the researcher left the room to run an errand and then have two. Years later, the children who 'delayed their gratification' scored better on SAT college entrance exams, were less likely to smoke, take drugs or become obese and were less physically aggressive. More recently, psychologists who followed a thousand people in New Zealand for decades discovered that those with the least self-control as young children grew up to exert a significantly greater burden on society as adults, with higher rates of unemployment, poor health and criminality.[3]

While there is some debate about the strength of the findings, there is a consensus that willpower and ability to delay gratification are skills that can allow us to make good choices elsewhere in our life despite temptations to do otherwise. For instance, in one study, smokers who were able to give up sweets – a classic test of willpower – were also more likely to quit cigarettes.

If you're one of those who think your willpower is naturally in short supply, the good news is that you can increase it. That's a relatively new insight. Older studies had suggested that willpower can be explained as a sort of fuel in our mental reserves: we only have so much of it, and expending too much leaves stocks temporarily depleted. That was based on studies where, for example, volunteers were asked to concentrate on solving a difficult maze, say, or a series of anagrams; their performance on a second mentally taxing task dropped compared with control volunteers whose initial task had been easy.

In another study, participants were given a freshly baked plate of biscuits and a bowl of radishes. Some were told they could eat what they wanted, and others that they could only eat the radishes. The volunteers were left alone for a few minutes to eat, before being given an unsolvable puzzle. Those who'd been told to lay off the biscuits, and therefore presumably used more willpower to resist temptation, gave up more easily on the puzzle. More than a hundred other studies have shown a similar pattern – tax our mental energy and our willpower diminishes.

Researchers went on to suggest that the currency of this effect is glucose, which gets used up faster when our brains work hard. Some researchers suggested eating regular snacks when doing anything mentally challenging – a catch-22 if the thing you're trying to resist is overeating.

However, more recent research indicates that our levels of willpower are not so much a budget that we have to eke out, but a renewable resource that can be powered up through the day, as long as you know how. The big secret behind that is actually pretty simple: your willpower is only limited if you think it is. This more nuanced picture emerged by combining willpower experiments with mindset studies. Previous work has shown that people with more malleable mindsets, who don't think their abilities are set in stone, are more resilient and persistent when it comes to their performance in a range of fields.

In this light, Stanford psychologist Carol Dweck and her colleagues performed willpower experiments, but this time asked volunteers beforehand whether they considered willpower to be a limited resource depleted by effort. They found that those who believed that willpower is finite showed the usual depletion effect when faced with a second challenging task, whereas those who believed that willpower is potentially unlimited showed no signs of running out of steam.[4]

In fact, subsequent studies discovered that you can improve willpower just by telling people that such a thing is possible. When volunteers were shown statements such as 'It is energising to be fully absorbed with a demanding mental task', they continued to improve through a tough twenty-minute memory challenge. Another group that was told willpower is limited stopped improving about halfway through the same task.[5] As with physical effort, in which our muscles feel tired long before they are close to collapse, how long we can keep going is all about how much energy we think is left. Which means that you've already improved your willpower, just by reading this paragraph.

If you need any more convincing, think about this experiment. In India, there is a widely held belief that mental effort isn't draining, but energising. Researchers have long suggested that this cultural attitude explains why Indian kids tend to spend more time in class and more time on homework and reading books than many in Western nations, with less concern about burnout.

When classic willpower depletion experiments were repeated on hundreds of Indian participants, they actually showed the reverse results — when the first task was harder, they tended to perform better on the second task. That all goes to say that willpower depletion is by no means an inevitable feature of human psychology. In fact, it could mean that seemingly super-human powers of willpower and self-control may be within anyone's reach, given the right mindset and a little practice. See mental challenges and resistance to temptation as energising rather than draining and your beliefs may well become self-fulfilling.

Not only will this help you on your way to resisting those

bad habits and forming new ones, but it might also improve other parts of your life. University students who believed that willpower is unlimited were not only happier, but suffered less from stress and bad moods when exam time approached.[6]

Oh, and there's one more thing you might want to do when you're looking for an extra ounce of self-control: hold off from visiting the bathroom. That at least was the conclusion of a study by Mirjam Tuk at the University of Twente in the Netherlands, which asked volunteers to either take a few sips of flavoured water or a few glasses before being made to wait a while, and then take part in numerous tests. These included a classic test of self-control, in which they had to consider whether they would prefer to receive a small amount of money now or a larger sum at a later date. The subjects who had downed the drinks, and were closer to needing the toilet, were more likely to choose to wait. It seems flexing our willpower in one domain might bolster our resolve in another.

One word of caution though. Too much willpower might be bad for the body and the mind. The trick is to know when to give in to temptation. In a series of economic games, researchers discovered that people who have greater willpower and self-control acted more selfishly than others, particularly if they felt their actions were private, without consequences for their reputation. In the real world, it's easy to imagine how this might be reflected in cases of fraud, for instance, which don't tend to be the result of immediate impulses but instead require steady dedication and organisation.

Indeed, it seems self-control can amplify your moral sensibilities, for better or worse. In one study, students had to complete two weeks of self-control training. They had to concentrate on using their non-dominant hand for everyday tasks such as opening doors – a simple intervention that had previously been shown to increase self-control. Their feelings of moral responsibility were also measured using a questionnaire.

Here comes the creepy part. The students were then called into the lab and given a vial of twenty cricket nymphs, along with a modified coffee grinder, which the team called the 'extermination machine'. The task was simple: feed the bugs into the grinder (in reality, the cricket nymphs crawled to safety through a concealed emergency exit). For people with a strong sense of moral responsibility, the lessons in self-control helped them resist the experiment's orders, compared with people with a similarly strong sense of moral responsibility who had not undergone the lessons. But the opposite was true for the more amoral students: those who had had lessons in self-control tipped about 50 per cent more nymphs into the grinder than people in a control group. Perhaps their greater self-control made them more obedient, or perhaps it allowed them to inhibit feelings of disgust as they fed the nymphs into the machine.

Whatever the reason, the findings indicate that self-control and willpower are more complex than we might think, both interacting with other personal traits to promote very different kinds of behaviours. What's more, people with high self-control report feeling less satisfied with their partners and colleagues, believing that others take advantage of their dependability. Maybe we are so used to seeing them quietly persevere, we forget the personal sacrifices they are making.

Whether it's a bursting bladder or positive affirmations, it seems a nuanced approach to boosting our willpower is necessary. Self-control is a tool to help us get what we want from life, and we need to learn to recognise when it would be wiser not to wield it. Sometimes you need to work on curbing your impulses to break a bad habit. At other times, just let your inner Cookie Monster prevail.

HOW TO STOP SMOKING

Now you've learned a little more about self-control it's time to put it into action. Some of you might have turned straight to this chapter to learn the quickest and easiest route to giving up your worst habit – smoking. Although the World Health Organization reports that the prevalence of smoking is declining worldwide, in 2017, 15 per cent of adults in England smoked, causing around 78,000 avoidable deaths per year. Each cigarette knocks about fifteen minutes off a regular smoker's lifespan.

Cutting down will help. The risk of lung cancer, for instance, increases linearly the more you smoke. But even social smokers put themselves at increased risk of several diseases – every cigarette you smoke or lungful of second-hand smoke you breathe in from others introduces fine particulate matter into your lungs, damaging their lining and potentially causing serious breathing disorders, as well as cardiovascular disease and heart attacks.

Quitting will undo a lot of this damage. You can claw back some of those years lost, and within a year of not smoking, your relative risk of smoking-related heart disease is halved. But how to do it? Public Health England say that the most effective way to quit smoking is to use your local stop smoking service, which provides support groups, behavioural therapy and pharmacological techniques. But only 16 per cent of people trying this route are successful in the short term.

For those tempted to apply the lessons of the previous section, the least successful method for quitting smoking is going cold turkey and toughing it out using willpower alone. No matter how much willpower you believe you have, less than 10 per cent of people who quit cigarettes do so without any additional help from nicotine-replacement products. A phased approach seems to make little difference: researchers comparing studies involving more than 22,000 people found the success rates of those who quit smoking abruptly on a designated day and those who reduced their smoking behaviour before quitting were much of a muchness.[7]

In recent years, swapping to e-cigarettes, or 'vaping', has become popular. E-cigarettes are handheld, battery-run devices that vaporise liquids typically containing nicotine, along with other chemicals and sometimes flavourings, but free from the tar found in cigarettes. These gadgets claim to deliver fewer harmful compounds to your lungs than a cigarette. The rationale is that the harm from smoking comes not from nicotine, but all the other compounds in tobacco smoke.

It sounds like a good exchange. Some 3.6 million people in the UK and more than 10 million in the United States are now vapers. But research around the devices is a complex and rapidly changing landscape that is often contradictory and difficult to keep track of. For instance, in 2019, in the United States there was a sharp rise in lung injury and deaths associated with e-cigarette use. The US Centers for Disease Control and Prevention (CDC) said at the time that the only way to ensure you aren't at risk is by refraining from using all e-cigarette products. But at the same time, UK health bodies disagreed, with the statement that vaping is '95 per cent safer than smoking' – taken from a report by Public Health England – widely repeated.

The problem is nuanced. Most e-cigarettes still emit numerous potentially toxic substances in addition to nicotine, including acrylonitrile, butadiene and acrolein, and there are no long-term studies of their effect on health. There's also the issue of illicit products. Among a sample of 867 people diagnosed with vaping-related lung injury, 86 per cent reported having vaped tetrahydrocannabinol, or THC, the chemical that gives cannabis users a high. These THC-containing e-liquids were probably obtained from the black market.

It's unlikely that THC itself is to blame for lung injury, however – we haven't seen the same symptoms in cannabis smokers, for instance. But other chemicals are often used with THC in e-cigarette liquids. The CDC has flagged vitamin E acetate, a synthetic form of the vitamin, as the most likely culprit. In a recent investigation, it was found in all the lung samples taken from twenty-nine people with vaping-related lung injury.

This might explain why there hasn't been such an upsurge in vaping-related lung injury in the UK – under EU law, many e-cigarette ingredients, including vitamins, are banned as a precaution. The difference may also come down to culture. A US survey found that young people who vape are twice as likely to become smokers, but the same trend hasn't been found in the UK, where it seems they are more likely to be used as an aid to stop smoking, rather than a trendy lifestyle product.

Even legally obtained vaping products aren't necessarily completely safe – it's difficult to say what chemicals in e-cigarette vapour are reaching the lungs. Studies suggest that many people who vape have markers of lung disease similar to those found in the lungs of people with emphysema, a condition that causes shortness of breath and shortens life expectancy. In addition, nicotine itself isn't completely benign, and may still cause cancer: when human lung and bladder cells are grown in the lab, they turn cancerous at a higher rate if exposed to nicotine.

Long-term epidemiological data on any potential link between e-cigarettes and cancer won't be available for another few decades or so.

For now, however, a lot of evidence suggests that at the very least vaping is a better option than smoking – which is admittedly a pretty low bar. Several major reports conclude that vaping is likely to be 'far less harmful' than conventional cigarettes. A recent study has also shown that long-term smokers who switched to vaping were halfway towards achieving the vascular health of a non-smoker within a month. It was the largest clinical trial to date of the potential benefits, showing those who ditched cigarettes and vaped instead saw their blood vessel function increase compared with those who continued smoking. Whether this benefit is sustained is not clear, but if it is, those who switched would have significantly reduced risk of cardiovascular events, such as heart attacks.[8]

Less harmful is great, but can e-cigarettes help people quit nicotine altogether? Until last year, evidence was limited regarding their effectiveness, compared with other nicotine replacement therapies. But a recent influential study has concluded that the use of e-cigarettes seems to work better than other forms of nicotine replacement therapy when it comes to quitting. Peter Hajek from Queen Mary University of London and his colleagues randomly assigned 886 people who wanted to give up smoking either nicotine-replacement products of their choice, provided for up to three months, or an e-cigarette starter pack, with a recommendation to purchase further e-liquids of the flavour and strength of their choice. They all received weekly behavioural support for at least four weeks to encourage their desire to quit. After one year, the participants were followed up and asked to give blood samples to validate whether they'd abstained from smoking or not. Of those in the e-cigarette group, 18 per cent had given up smoking, compared with 9.9 per cent of the nicotine-replacement group.[9]

There are other, more benign, ways of helping you to quit – giving yourself a financial incentive, for instance. In one study, volunteers had to deposit $150 of their own money, with the promise that they'd get it back, plus $650 more, if they had verifiably quit smoking after a year. Of these people, 52 per cent quit. In another group, with an $800 incentive but without having to deposit any money, only 17 per cent were successful, suggesting personal financial loss is a strong motivator for good behaviour.

Personal incentives of a different kind can work wonders too, in my experience. I remember one hot, summer's evening, full of too much wine, telling my twenty-a-day dad that I just wanted him to still be alive to walk me down the aisle. According to my mum, later that evening he threw away his packet of cigarettes. Fifteen years later, he has never smoked again.

Then there's Allen Carr. In the 1980s, the UK self-help author made a huge impact and created a multi-million-pound empire around quitting smoking. His method didn't stop you from smoking, not immediately at least. In fact, you were encouraged to continue until you fully understood the 'nicotine trap', which was the realisation that the relief smokers feel from drawing on a cigarette was what non-smokers feel the entire time. Carr's 'Easyway' clinics were set up around the world and he published several books, starting with the 1985 bestseller *The Easy Way to Stop Smoking*.

It's pretty incredible that it took thirty-five years for anyone to test his method against others in a randomised controlled trial, the gold standard for this sort of thing. Finally, in 2019, a total of 620 participants who all wanted to stop smoking were randomly split into two groups. One followed the Easyway method focusing on the idea that smoking provided no benefits, which included a day's session of group-based support, alongside text-message support and a top-up session if needed. The second group attended a stop smoking service that provided behavioural and drug support in accordance with UK national guidelines.

The participants checked in six months after the day they stopped smoking and their self-reported success (or not) was verified using breath carbon monoxide measurements. Of those who attended all the treatment, 19.4 per cent of the Allen Carr intervention stopped smoking compared with 14.8 per cent in the standard intervention. This seems like a win for Carr, but when you account for the vagaries of statistics there was no difference in the efficacies of the two methods.[10]

What is important to understand is what works for one person isn't necessarily going to work for another. It will depend on your personal circumstances, personality and even your genes: a study of more than 11,000 people showed that two-thirds of white smokers carried a variation of the gene that made them about 22 per cent more likely to be able to quit smoking than those who carried alternative versions of the gene.[11]

Just remember what we learned earlier in the chapter about forming and breaking habits more generally: whatever route you choose, a few slip-ups are not the end of the world. Keep trying, get support, believe in your desire to do it and you will succeed eventually. In the end, a positive mindset may be the difference between failure and a healthier, smoke-free future.

HOW TO DRINK SENSIBLY

Some experts believe that the world's first breweries might have been created when grain stores became drenched with rain and warmed in the sun. Ever since, humans have discovered that alcohol reduces their inhibitions, impairs their judgement, influences sexual desire and performance, creates beer bellies and leads to hangovers (and dubious cures for hangovers).

What's in no doubt, either, is that when consumed in excess alcohol can lead to a slew of health problems, including but not limited to liver disease, brain damage, infertility and at least seven different kinds of cancer. The WHO goes so far as to class it as a group 1 carcinogen, because of a strong link between alcohol consumption and cancer, and says there is no safe level of consumption. And, of course, excessive alcohol consumption increases the risk of death from accidents and violence.

While there seems to be a downward trend in drinking among younger people in the UK, we may expect to see fluctuations in this pattern as a result of the coronavirus pandemic and its economic fallout. Deaths from liver disease rose in the United States after the 2008 financial crisis, with a particularly sharp rise in alcohol-related cirrhosis among young people, a reverse of the previous decade where deaths from cirrhosis were falling. Because of the timing of the upswing in deaths, the researchers suspected it could be connected to unemployment and other economic factors resulting from the financial crisis. Deaths from suicide and opioid use also rose during the same period.

As with smoking, when it comes to making and breaking an alcohol habit your genes might be working against you. For instance, a gene called ADH is involved in metabolising alcohol. ADH makes an enzyme called alcohol dehydrogenase, which breaks down alcohol into acetaldehyde, a toxic substance that is then transformed into harmless acetate by another enzyme. Some variants of ADH seem to protect against alcoholism, perhaps by breaking down alcohol so quickly that the toxic acetaldehyde accumulates in the body faster than the body can clear it. A person who metabolises alcohol like this would feel unwell even after small amounts of booze, and would be unlikely to drink enough to form any kind of alcohol addiction.[12]

Rightly or wrongly, most of us would not classify our relationship with alcohol as an addiction. I must say at this point that, if you are seriously concerned about an inability to control the amount you are drinking, or the impact of your drinking on yourself or those around you, then put this book down now and seek professional advice, either from your family doctor or from local and national alcoholism advice services, for example Alcoholics Anonymous and Drinkline in the UK. Treatment often includes counselling, cognitive behavioural therapy and drug therapies. The most common prescription is chlordiazepoxide, sold under the name Librium among others, which can help you cut down on alcohol while minimising withdrawal symptoms. You might also be offered acamprosate or naltrexone, which reduce any urge you have to drink over the course of a year.

When it comes to 'moderate' drinking, the sort most of us would happily admit to – say one or maybe two drinks a day – the health benefits, or otherwise, represent one of the most hotly contested subjects in nutritional science. Many people think that a few glasses of red wine, in particular, can protect you from things like heart attack and stroke. And hundreds of studies seem to back this up, to greater or lesser degrees. Red wine gets a particularly good review because it's packed with resveratrol, a plant chemical that in animals can protect against some of the effects of stress and malnutrition. It has also been shown to be a powerful anti-inflammatory and an antioxidant, as well as showing signs – mostly in animal studies, but a few early human trials, too – of protecting against arthritis. In humans, a pill that contains resveratrol has also shown promise in slowing the symptoms of dementia. The problem with equating this to drinking more red wine is that to imbibe as much resveratrol as the drug, you'd have to down a thousand bottles a day.

If, like me, you like to think that a glass or two can help you relax, and de-stress, and therefore might be helpful in countering some of the negative consequences associated with stress, then you're not alone. Unfortunately, there's not a lot of evidence to back this up. If moderate drinking were a major factor in protecting against stress-related illnesses, it probably would have been picked up in epidemiological studies by now.

A major problem with research into this sort of thing is that people who drink in moderation also tend to live healthier lifestyles in general – they don't smoke as much, they are more affluent, they eat healthier diets and exercise more. It could be these things that are protecting them against heart attack and stroke, rather than the few glasses of wine they crack into each week. Other confounding factors are that most studies don't differentiate in their control group between those who have never drunk alcohol and those who might have given up drinking because they have already done themselves damage through too much booze. Their health problems may disguise the benefits of an alcohol-free existence. All this means there is little concrete scientific consensus on any benefits of alcohol consumption, I'm afraid.

The UK's current guidance recommends you drink no more

than fourteen units a week – the equivalent of six pints of regular-strength lager – spread over at least three days. It no longer distinguishes between men and women, even though alcohol does affect men and women differently. Women have a higher blood-alcohol content than men when they drink the same amount, but metabolise it faster, meaning it comes out of their body quicker. Women are at higher risk of some alcohol-related cancers, but men are more often heavier drinkers and are more likely to partake in risky behaviour and violence, so the two appear to balance out in terms of harm.

If you really want a sobering analysis of the risks, a review of nearly 600,000 people in nineteen countries found that anyone drinking more than around five 175ml glasses of wine or five pints of beer a week was at an increased risk of early death. But to give this some context: 25,000 people would need to drink an additional ten extra grams of alcohol a day – a little bit more than a half pint of normal strength lager or a single shot of spirit – for one of them to develop a drink-related disease each year. Given that everything we do comes with risk – eating cake, driving a car – you may decide that this one is worth it.

You might decide that the best balance between benefit and harm lies with cutting back on alcohol, rather than giving it up completely. Everything you have learned so far in this chapter will help, particularly identifying and avoiding the cues and environments that your brain now strongly associates with a drink or two.

Another increasingly popular strategy is partaking in 'Dry January', or 'Sober for October'. The idea is that you get on the wagon for a month, helping to give your body a break from alcohol, time to recover and space to reset the relationship you have with it.

But does a temporary period of abstinence do any good? Because it's a relatively new societal phenomenon, studies have only just started. Some of my colleagues at *New Scientist*, having posed the question, participated in one of the first trials a couple of years ago. The results from these tests look promising. In the short term, a month of zero booze reduces blood pressure, cholesterol and two growth factors associated with cancer. It also reduces insulin resistance, which suggests an associated lowering of risk of developing type 2 diabetes.

Whether a month off the booze makes a difference in the long term is still in question. We don't know, for instance, whether people who quit for a month drink more the rest of the year as a consequence. It may be better to go 'damp' rather than dry – making sure you have at least two days a week on which you don't drink throughout the whole year, rather than stopping for a month and then thinking you can do whatever you like the rest of the time.

Something that I've found particularly helpful when it comes to controlling alcohol consumption and other such vices is to think about my microlives. This is an ingenious way of over-coming the psychological block we all have when we consider how our choices today might affect us further down the line. We all tend to value the short-term gain of an indulgence such as a drink today, over its potential long-term effects, such as an increased risk of liver disease or cancer. One microlife is a millionth of a life, and is the equivalent of about half an hour. Every unit of alcohol you drink today knocks half a microlife off your total. So too does smoking one cigarette or sitting in front of the TV for an hour.

You can use microlives the other way too – twenty minutes of physical activity adds one microlife, for instance. So if you do decide to have that extra pint of lager, with its two units of alcohol, think about whether it's worth the effort of working out for an extra twenty minutes to balance the books.

HOW TO STOP OVEREATING

If you feel like your diet has completely gone to pot during a year of coronavirus lockdowns in which you've struggled to find your regular food, and been stressed and bored – known risk factors for overeating – then you're not alone. Emerging evidence suggests that many people, in the UK at least, are struggling to resist the comforts of food more than ever. Weight gain and its negative impact on our health might be an unforeseen consequence of the coronavirus pandemic.

An unhealthy diet is often blamed on poor choices and a lack of willpower. We've already looked in Chapter 6 at some ways you can train your brain into choosing healthier foods. But new research reveals another line of attack specifically if you find that you are consistently overeating. This may be due to the fact that you have several different appetites. Understanding how they interact could be key to cutting back on your overindulgences.

Let me explain. The idea started with Stella. She lives on the outskirts of Cape Town, a beautiful, rural setting surrounded by vineyards, trees and wild heathland. In 2010, researchers at City University of New York followed Stella for thirty consecutive days, recording exactly what, and how much, she ate.

Stella's diet was extremely diverse: she ate ninety different foodstuffs over that time, with her ratio of fats to carbohydrates varying wildly day to day. But when the team crunched the numbers, looking at the ratio of combined daily calories from carbs and fats to calories from protein, they always got close to

4:1, regardless of what she ate. It was surprising because this ratio is nutritionally ideal for a female of Stella's size. In that sense, she was a meticulous eater. And the question was how she was getting it regularly so spot on, when even professional dieticians have trouble, turning to computer programs to help. Stella didn't have access to this kind of technology, because she was a wild Cape baboon.

This is just one of many studies over the past thirty years that have changed our understanding about appetite. When we think about our own appetite, we tend to think about it as one thing. We're hungry or not. Or we're craving one particular foodstuff or sated. It was a couple of decades before Stella, in 1991, that researchers began to study how we choose what to eat when we're hungry.

Back then, they started with locusts, putting hundreds of them in individual boxes and preparing twenty-five different foods containing various proportions of protein and carbohydrates, the main nutrients the insects eat. The foods ranged from high-protein/low-carb to high-carb/low-protein, and everything in between. Each locust was fed just one of the twenty-five formulations, in unlimited quantities, until they reached adulthood. The team meticulously recorded how much each locust consumed each day, plus their weight and how much fat and lean tissue they had put on.

The ideal diet for locusts to thrive turned out to be around 300 milligrams of carbs and 210 milligrams of protein a day. What was striking is that all of the locusts managed to get close to the ideal amount of protein, even if that meant missing the carbs target by miles. Those on a low-protein diet thus ate a huge amount of carbs. This came at a cost of getting fat, like an overweight knight wedged into a small suit of exoskeleton armour. In contrast, the locusts on a high-protein diet ate too few carbs and were unhealthily lean. They were less likely to reach adulthood, and those that did had too little body fat to survive in the wild.

It was a battle between protein and carbs: when the locusts' food didn't allow them to eat a balanced diet, they prioritised protein over carbs at great cost to growth and survival. In fact, it wasn't so much a competition between nutrients as between two separate appetites – one for protein, the other for carbs. Just like Stella, when the locusts are given a wide choice of foods their two appetites collaborate so they consume an optimal diet. But when they are given imbalanced foods, the appetites for protein and carbohydrates compete, and protein wins. That suggested that, more so than carbohydrates, protein has to be carefully calibrated in the diet, probably because it helps animals grow and reproduce. We now know it's a similar story in organisms from slime moulds to spiders, cats and dogs.

But what about humans? A pilot study took ten friends and family to the Swiss Alps, where they could choose whatever they wanted to eat from a highly varied buffet for two days. Then, on days three and four, half the volunteers got offered a high-protein buffet, and the other half a low-protein, high-carb and high-fat buffet. For the final two days, they returned to the original diet.

In phase 1 of the experiment, the human locusts reliably got about 18 per cent of their calories from protein, in keeping with studies that show people typically need 15 to 20 per cent. In phase 2, everyone maintained their absolute protein intake – meaning those on the low-protein diet had to eat 35 per cent more total calories, while those assigned the high-protein diet ate 38 per cent fewer calories. The humans responded like locusts, with their appetite for protein dominating, and determining the total consumption of food.

Two more sophisticated versions of the chalet experiment in Sydney and Jamaica have confirmed the result. More recent findings suggest that we in fact have five appetites that all keep track of a specific nutrient, with the aim of constructing a balanced diet: for protein, carbs, fats, sodium and calcium. It's likely that these five foodstuffs have been singled out by evolution because they are needed in very specific quantities. Some, like sodium, were rare in our ancestral environments so we needed dedicated machinery to seek them out. By focusing on these five, we also tend to eat the right amounts of other essential vitamins and minerals.

The startling conclusion from this dance of the five appetites means that in a food environment that is protein-poor but energy-rich, people will overeat carbs and fats as they strive to reach their protein target. This is exactly the environment those of us living in rich-world economies are exposed to. According to the UN's Food and Agriculture Organization, between 1961 and 2000 the proportion of protein in the average American diet fell from 14 per cent to 12.5 per cent, with the balance made up of fats and carbs. Given that shift, the only way people could have maintained their target protein consumption was to increase total calorie intake by 13 per cent – more than enough to create an obesity epidemic.

Intriguingly, the experiments above showed that the extra calories eaten by those on a low-protein diet came from savoury snacks, especially those that tasted like umami, the signature flavour of protein. The protein-deprived volunteers were craving things that tasted like protein, even though they were made of carbs. Our food environment is awash with these kinds of protein decoys – crisps, instant noodles, crackers and so on. They are also known as ultra-processed foods.

It's no surprise that ultra-processed foods, designed by industry to be irresistible, are bad for us. They include delicious common fare such as pizzas, sweets, bread, cakes, mayonnaise, ketchup and ice cream. But it might be that the problem of overconsumption has less to do with these foods being full of fat and carbs, as is often depicted, and more that they are depleted of protein.

They tend to be low in protein because it's an expensive addition. But when protein is diluted by fats and carbs our appetite for it overwhelms the mechanisms that normally tell us to stop eating fats and carbs. Ultra-processed foods also contain very little fibre, which is filling and so puts a brake on appetite. Their frequent flavouring with umami, which our protein appetite craves, only makes matters worse. As a result, we eat way more than we should.

The good news is that you can use these insights to make your five appetites work for you, rather than against you. The first step is to calculate your protein target. To do so, first look up the daily energy requirement for your age, sex and level of activity. You can do this with something called the Harris-Benedict equation calculator, available on numerous websites. Mine, as a 5-foot-10-inch, 36-year-old moderate exerciser, is around 2,200 kilocalories a day.

Next, work out the portion of those calories that should come from protein by multiplying by the proportion of energy in our diet you should get from protein; this multiplier varies depending on age: for eighteen to thirty-year-olds it is 0.18 (indicating 18 per cent of all energy should come from protein), for people in their thirties it is 0.17, for those aged forty to sixty-five around 0.15 and for those over sixty-five 0.2.

Divide the resulting number by 4 to get the number of grams of protein per day you should eat (a gram of protein contains 4 kilocalories of energy). Mine comes to between 84 and 94 grams, depending on how much exercise I do each week.

Finally, work out how to obtain that from protein-rich foods such as meat, fish, eggs, dairy, pulses, nuts and seeds. This is slightly complex, but the protein content of all these foods is available online and on food labels. I could eat, say, one large egg (6g) at breakfast, a handful of walnuts (4.5g) for a snack, a salmon fillet salad (40g) with avocado (4g) and lentils (9g) for my lunch, and a fillet steak at night (24g) to take me up to my total.

Of course, how you choose to consume your protein will make a difference to your overall health. Red meat, for instance, might be full of protein but is also high in saturated fat, and has been linked to an increase risk of heart attack and cancer. But the idea is that by focusing on your protein consumption, everything else can flow from this. Satisfy your protein appetite and it will automatically ensure that you don't overeat on carbs and fats. In theory, you won't need to keep track of your carbs and fats at all as your protein appetite will manage them for you. Just make sure you supplement the high-protein foods with whole foods – plant-based items that are processed and refined as little as possible, such as legumes, fruits, vegetables, rice and wholegrain cereals.

Most important, avoid ultra-processed foods. Keep them out of the house. You will eat them if they are there. They are designed to be irresistible. If you follow these steps, the rest should be easy. All you have to do is listen to your appetites: they will guide you towards a healthy and satisfying diet. That is what they evolved for: to work for you, not for food companies.

HOW TO DIGITALLY DETOX

'You'll get square eyes!' I found myself saying to my daughter the other day. My mum used to say it to me too. These days it's hard to avoid – even if you're not in front of your phone, we all encounter screens on the kitchen table, on the sofa, by the bed – they're constantly accessible in a way that my mum couldn't have imagined all those decades ago. If you believe the headlines, screens are supposed to warp our skeletons, damage our mental health and alienate us from our families. But do they? And if so, what can we do about it?

Where once the bulk of our addictions were to *things* like cigarettes and alcohol, now in the developed world they seem to be to *behaviours* – posting on Facebook and Instagram, binge-viewing Netflix, or obsessively checking emails, all things you can do with that little gadget in your pocket, your smartphone. When we look at smartphone use, we can see that our constant notifications tend to elicit a positive emotional response, triggering those familiar pleasure centres in the brain.

Even so, smartphone use probably doesn't actually count as an addiction for most of us, if you go by the usual definition. Few people so obsessively check Twitter that it ruins their life, or commit crimes to feed their Instagram habit. But while the reward might not be as strong as substance or gambling addiction, if you're one of the 2 billion people around the world who is attached to their smartphone, finding yourself obsessively checking your phone every time you feel a vibration or hear an alert, you may have a terrible habit. On average, you probably touch your phone around 2,600 times a day. Imagine how many hours that adds up to over the year.

Our smartphone habit really became apparent in the days of

Tetris, shortly followed by Candy Crush – viral games that people couldn't stop playing. Candy Crush in particular became an instant, unstoppable juggernaut and pop-culture phenomenon. One anonymous web confessor told how when she finally got off the toilet after four hours of play her legs gave way beneath her.

Many of these games tap into psychological techniques that keep us hooked. Rearranging shapes is undeniably deeply satisfying. It appears to trigger something called the 'ludic loop', a tight, pleasurable feedback loop that stimulates repetitive, if not compulsive, behaviour. Slot machines illustrate this concept perfectly. They lure people into short cycles of repeated actions using lights, singing sounds and occasional cash rewards. You do it again. And again and again. Our affinity for this kind of activity comes back to dopamine. Although it's often thought of as a simple reward or pleasure chemical, it's also linked to the compulsion to repeat an activity, whether or not that activity is pleasurable.

That is why slot machines, which beget compulsive behaviour despite offering virtually no chance of a tangible long-term reward, still appeal. There's no goal, just the pleasure of being in this zone – the ludic loop offers the brain its own reward. Games like Candy Crush hit upon this formula, too, a kind of sweet spot between pure chance and the illusion of control with the occasional reward. Random windfalls are much better at making us compulsively repeat certain behaviours than predictable ones. That is why we also crave looking at our phone when we hear that random buzz – even though it could just as easily be an alert to pay your credit card bill as a message from a new lover.

And while the games may have passed their peak, there's still plenty to keep us hooked to our screens. Now, however, you're more likely to have sat down to watch one video on YouTube and woken up two hours later with no idea how so much time passed. More than 70 per cent of viewing time on YouTube consists of people watching videos suggested by the platform rather than sought out deliberately. We have artificial intelligence algorithms to thank for that – they learn what our brains find pleasurable because of all the data we leave behind online. Apps such as Facebook, Twitter and Instagram all use tricks to command our attention. Think about the satisfaction of pulling down on the screen to refresh it – it shares characteristics once again with the mechanism of slot machines, with the same addictive qualities.

When you think about the jolts of dopamine that we're getting from our digital devices, it's really not surprising that we are liable to walk out into the road and into a passing car because our nose is in our phone. But aside from the danger of not paying attention to the road, or wasting time watching videos of cats and TikTok dances, is this habit really damaging? Is my daughter really going to get square eyes?

One problem with satisfactorily answering these questions is that 'screen time' covers a multitude of purposes. We might be using it to bank, to create photo albums, to chat with colleagues. We use screens for work and play, to record physical activity, to monitor sleep. We can look up peer-reviewed papers or WhatsApp our friends – the umbrella term 'screen time' no longer has a distinct definition. So despite many studies, there's no scientific consensus as to whether it's damaging or not. The World Health Organization recommends limiting screen time as a way of tackling obesity because as we know, lazing around is a fast track to putting on weight and putting yourself at risk of several diseases. But that's not a health concern related to screens in particular.

It's true that in the past few years more children in the UK have been prescribed glasses, but this might have more to do with aggressive management of existing conditions. There's not sufficient evidence for a causal link between watching too much Peppa Pig and damaging your kids' eyesight.

One thing that is taking a hit is sleep. People who read a book in bed find it harder to go to sleep if they read it on a screen rather than on paper, because of the blue light that most screens emit, which throws off our circadian rhythms (if you missed it, see Chapter 7 on sleep for more details on that). Most of us aren't reading books on our phones before bed, though. If you're me, you're getting worked up about ignorant opinions when scrolling through Twitter or planning your dream home from an Instagram feed. But whatever we're doing, as we know, insufficient sleep is linked to all kinds of health problems. If there are any recommendations concerning screen time to be adopted, turning off your screens, or at least limiting screen time, in the hours immediately before bed is probably the one with the most support.

But what about the fact that our digital companions are tapping into our need for social validation, hooking us on likes, retweets and follower counts? In 2019, Tristan Harris, a former Google designer and co-founder of the Center for Humane Technology, told the US Senate that the internet has created a culture of mass narcissism. Is it true? Have we wrecked a generation of adolescents, priming them for increased risk of mental health disorders, depression, anorexia and suicide?

Dozens of papers would suggest so. Harris pointed out in his testimony that mental health problems in ten to fourteen-year-old girls was in decline for around two decades, but has shot up 170 per cent in the last eight years. The problem with these findings is that the underlying data can be used to tell a totally different story. Social media is just one of many different things that might affect someone's well-being, and without controlled studies it's difficult to draw meaningful conclusions.

When Amy Orben at the University of Oxford investigated screen use a few years ago she explored some of the most extreme claims researchers were making, such as linking social media use with teenage depression and suicide. To get some perspective on the matter, she and her colleagues compared the effect of device use to other things in an adolescent's life. For instance, they looked at the effect of wearing glasses and found that this was correlated more negatively with well-being than screen use. They also looked at how often adolescents ate potatoes. That turned out to have a similar negative correlation to well-being as screen use – but that doesn't necessarily mean potatoes should be banned from schools.

The debate could go on, with other researchers considering flaws in Orben's own statistical methods. What seems to be true is that we cannot talk about social media for individuals as good or bad. Blanket statements just can't be supported by evidence. If you're confused about where this leaves us, join the club. Of course, we could turn to organisations such as the American Academy of Pediatrics (AAP), which discourages parents from allowing children under two to have any interaction with screens, and recommends no more than an hour a day for two to five-year-olds. But even here, the message is mixed. The World Health Organization says that children under three should have no screen time and those aged three to four should be limited to an hour a day, but its focus is on curbing childhood obesity.

If this fills you with horror, because you're used to occasionally distracting your child with some screen time when you have other things to do, you're not alone. Like me, you might prefer to stick with the UK Royal College of Paediatrics and Child Health's view, which has opted not to recommend time limits at all. They concluded that there isn't enough evidence of positive or negative effects for any guidelines to be issued.

It's a refreshing and honest way of creating evidence-based policy. The Royal College's view is that advice is only as good as it is effective, and any parent who supports an outright ban on screen time for children under two lasts until they spend a day where there are children of different ages in the same room, or have a week with a child who is unwell and they have a virus themselves. Then they realise it's unworkable.

And it's not all negative. In 2016, researchers found no evidence that spending time interacting with a screen – rather than moving around or interacting with other humans – delayed certain developmental milestones, such as learning to walk and talk. On the contrary, they found a correlation between screen use and earlier development of fine motor skills, such as the ability to pick up blocks and stack them in a tower.

Once again, however, there is no sure-fire causal link. While it is possible that the prodding and swiping needed to work a screen trains these skills, it could be that those infants who happen to develop fine motor skills early are simply more likely to pick up and play with a screen. We shouldn't discount the value of screen use for older children, either. Not only does it provide unprecedented access to many forms of valuable information and entertainment, but educating children about the dangers they will find online requires them to have some familiarity with it.

These devices that rule our lives are both maddening and valuable. Rather than worry about arbitrary constraints, perhaps the best advice is to look at our use of screens and ask how it fits with the activities and lifestyles we want as individuals and families. If you feel you need to gain a bit more control, apps that show you how many times you look at your phone could be a good place to start. Deleting alerts and notifications is also a good idea. We know these disrupt our work and lifestyle, so making a more conscious decision when to engage with these things can have valuable results. Writing this book, I took this advice and stopped all notifications appearing on my screen. This is just my anecdotal evidence, of course, but I felt it improved my well-being more than I could have possibly imagined. Now, I only look at my emails during work hours. I have to choose to interact with WhatsApp and I rarely get lost in YouTube videos, unless I've specifically given myself time to do so.

If this doesn't sound like you, introduce a 'mental speed bump' that interrupts the habit of checking your phone, such as writing a note to yourself on your lock screen or simply wrapping a rubber band around the device as a reminder. Or use night-time settings and wind down modes designed to make the screen less enticing by turning it black and white. Whether this results in positive health and lifestyle changes is still an unanswered question. Every new technology has given rise to new fears. An article from 1941 laments how adolescents in the United States were addicted to radio programmes.

Given the information you've collected from previous chapters in this book, you should have gained a bit more knowledge of what you should be eating, how much time you should be exercising, and what kinds of social activities might make you happiest. Ask yourself how these ideals fit in with your screen use, and if you find yourself overindulging, don't panic. Just think of what level of screen use makes you and those around you happiest and healthiest and try to stick to it.

TOP TIPS FOR BREAKING BAD HABITS

🔧 Remember that your willpower is not eroded over time – it's a potentially limitless resource as long as you believe in it. Just knowing this fact will help you perform better for longer.

🔧 The occasional slip-up shouldn't affect your progress, so don't beat yourself up about it. Remember that it takes time to break a habit – between eighteen days and close to a year.

🔧 It may be better to cut down on alcohol, giving yourself two alcohol-free days a week, rather than take a dry month every year. To help you cut back, think about your microlives, a unit equivalent to half an hour: every unit of alcohol you drink now takes away half a microlife. Twenty minutes of exercise adds one back on. You do the maths.

🔧 Stop overeating by calculating how much protein you need each day and making sure you get this. It will stop you trying to balance your protein intake using foods that mimic the taste of protein, but that are full of fat and carbs.

🔧 Cut back on your smartphone cravings by removing notifications, and making sure you limit the amount of time you stare at a screen before bed.

9

HOW TO BE SMARTER

THERE'S A STORY about me when I was a young kid that my mum loved to tell again and again. It was a Tuesday morning and I was ill and miserable. I shouldn't have gone to nursery, but the fire truck was paying us a visit and I wanted to see it. I was around three at the time and have vague memories of clinging on to my mum's leg as the older kids jostled to get in line to spend time in the driver's seat. The head firefighter said the first person to answer an important question could ring the siren. An important question? My ears pricked up.

'What number should you ring if you have a fire?' he asked. Around me, friends' hands shot up. 'A hundred!', 'One two three!', 'Four four four!' Kids shouted out random numbers from all corners. Finally silence. As miserable as I was I just couldn't bear it. I sighed. 'Nine. Nine. Nine,' I said in a voice (according to my mum) that sounded like I thought the rest of my class were utter morons.

I don't know why this story got repeated so often. But it did summarise a lot of my early life – I always enjoyed the feeling of knowing the right answer, of getting top marks, of pushing myself to be cleverer than others. I raced my best friend through every maths book my teacher could provide. I spent my pocket money on things like *The Guinness Book of World Records* and poetry pamphlets so that I could memorise facts and rhymes to recite at the next family party (what a blast I must have been). When I learned age seven that my nan, who was blind, could read using Braille, I sat for hours teaching myself the same skill and could soon read a whole fairy tale by touch alone. A few years later, I got into a super-selective grammar school with 100 per cent on a non-verbal reasoning intelligence test. I was smart, and I knew it.

What I didn't know, however, was what that meant – and what it didn't mean. That's why this chapter and the next one come in kind of a double pack. In this chapter, we'll discuss intelligence of the sort that can be measured by IQ tests. We'll look at ways we can boost both this functional intelligence and memory, and keep them functioning well into old age.

But in the following chapter, 'How to be wiser', we cast the net a little wider. Here we look at how the sort of intelligence I showed in abundance as a kid, the sort measured by IQ tests, is just part of a bigger picture embracing 'softer' aspects such as wisdom, emotional awareness and self-knowledge. It's worth bearing that thought in mind as you proceed through this chapter. But first, back to basics.

HOW TO UNDERSTAND INTELLIGENCE

Intelligence has enabled us to land on the moon, cure disease and generally dominate this small blue dot of a planet. But arriving at a universally agreed definition of what intelligence actually is still defeats us. It's also a subject that people often find uncomfortable to discuss. Often, that's because of a belief that intelligence is something you are born with and therefore can do nothing about. This undercuts social equality, and feeds into the link between intelligence testing and eugenics, an issue that still looms large for many.

We might try to broadly define intelligence as our ability to comprehend our surroundings, to catch on, to make sense of things or work out what to do. It's the ability to understand and learn from experience and change our behaviour accordingly. OK, so there is a stab at a definition – now how do we measure it? The most famous – or infamous, depending on your point of view – method is via 'intelligence quotient', or IQ. This relies on tests of verbal reasoning and of pattern-finding, the non-verbal reasoning I apparently excelled at.

IQ therefore fails to embrace the rich complexity of attributes that help us to behave intelligently: things such as wisdom, social sensitivity and practical sense. But although IQ tests often come in for criticism for their narrow focus, they are still arguably one of the most reliable predictive tests based on what we might call intelligence. (If you want to take one, don't be fooled, incidentally, by the sort of IQ tests you see advertised on Facebook: real IQ tests take at least an hour to perform, contain a mix of verbal, non-verbal, memory, mathematical and picture-based problems, and are often marked by trained professionals.) The tests capture many of the mental skills that correlate with performance in academic exams, appearing to account for roughly two thirds of the variance in school exam scores.[1]

Your IQ can also predict how you respond to workplace training and how well you do in your job, even in non-academic professions such as car mechanic or carpenter. It also predicts social mobility, perhaps because it reflects a person's ability to handle complexity in everyday affairs.[2] Many tasks, from grocery shopping to juggling our diaries, require us to deal with unexpected situations, to reason and make judgements and to identify and solve problems – many of the aspects that IQ tests examine. This is true of our social interactions too.

People who score better in IQ tests are also healthier and live longer. One explanation could be that they are more likely to be better educated, and therefore also more likely to be in jobs that command higher salaries, helping them afford things like gym memberships and healthier foods. Another is that learning, reasoning and problem-solving skills are useful in avoiding accidents, preventing chronic disease and sticking to complex treatment regimes if you do fall ill. In addition, a lower IQ might be caused by events during foetal development or childhood – such as a blow to the head – that influence health and longevity. All of that adds up to IQ being about the best measure of intelligence we have. That leads us back to the question, can we increase it?

There is no escaping the fact that intelligence is inherited to some degree. Long ago, studies showed that identical twins had very similar IQs in adulthood, which strongly correlated with that of their biological parents, even when they had been separated and brought up in very different families. Hundreds of studies have since pointed to the idea that about 50 per cent of the difference in intelligence between people is due to genetics.

However, while genes matter, they are certainly not destiny. While the twin data is valid, its interpretation that our IQ is immovable is not. At the heart of this idea is that people are drawn to environments that suit their genes: in other words, genes select environments. If, say, you happen to be tall then you are more likely to make it on to the basketball team and to receive coaching, play in competitive games and find your basketball skills racing ahead of a friend who is a little shorter. The new environment you have moved into amplifies a small genetic difference.

We can think of intelligence in the same way. A child who has natural cognitive skills may delight in more cognitively difficult tasks and make friends with those who feel the same.

The brain is more like a muscle that can get stronger in a workout than suggested by the simple 'genes make us who we are' view. Or to quote James Flynn, one of the most famous intelligence researchers in the world, 'Genetic differences between individuals appear dominant only because they have hitched powerful environmental factors to their star.'[3] What this means is that genes give us a blueprint that may influence our behaviour and set limits on our intelligence. Meanwhile our environment determines where within these limits a person develops. What that means for how we develop our intelligence is what we'll come on to next.

HOW TO BOOST YOUR IQ

We have evidence of all sorts of ways that making environmental changes affects intelligence, good and bad. For instance, iodine deficiency during childhood is associated with lower IQ and addressing this in developing countries has boosted cognitive skills. So too has treating parasitic worms and removing lead from petrol across the world. Meanwhile a demanding job can boost IQ, but early retirement may send it plunging. It points to one message: you can boost your brainpower – or dumb it down – throughout your life.

Perhaps the most well-documented intelligence-booster is more education. It's been shown to work time and again. True, intelligent children often remain in school for longer, but that isn't the whole story. During the 1960s, the Norwegian government added two extra years of compulsory education to its curriculum and rolled out the change gradually, allowing comparisons between different regions. When researchers investigated IQ scores from tests taken by all Norwegian men as part of their compulsory military service, they concluded that the additional schooling added 3.7 IQ points per year. This pattern has been seen elsewhere. One meta-analysis concluded that each additional year of schooling boosted IQ by between 1 and 5 points.[4]

Why that should be so is up for debate. It might simply be that reading, studying arithmetic, and accruing general knowledge are good training for the kind of abstract thinking you need to perform well in IQ tests, or that schooling teaches children to maintain their concentration, or something else entirely. Whatever the answer is, a good childhood education is good for your intelligence – and good for your life.

Whether adult education has a similar effect is less clear. It is a plausible hypothesis, although it hasn't been tested directly. But it's worth remembering that not all learning is in the classroom. One study compared people's IQ scores at the ages of eleven and seventy, and found that being in a more complex job was related to being smarter in later life – even after controlling for how smart a person was to begin with.[5] This group still had some age-related decline, but it was less pronounced than in other people. The evidence is consistent with the 'use it or lose it' hypothesis – that keeping our grey cells active is the best way to ensure they remain in good form. It points to things you can do to give your IQ a boost, even if your formal education is long past.

It would be intelligent to sound a strong word of caution here. During the early 1990s a paper was published in *Nature* revealing that students performed better on a test of intelligence involving spatial reasoning if they listened to Mozart while taking it.[6] So was born an entire brain-training industry.

Sadly, other researchers have been unable to replicate the

Mozart effect, and other related claimed effects. In 2007, the German research ministry commissioned a report analysing all the scientific literature on music and intelligence, and came to the conclusion that passively listening to music did not make you more intelligent. Studies of computer games that claim to improve mental performance have produced mixed results too. Brain-training, Baby Einstein and so on have all been fairly disappointing in terms of being able to boost IQ.

Recent studies suggest that the efficacy of brain-training is limited to small improvements in the types of memory or verbal skills you're practising in the brain-training tasks themselves. But these benefits only extend to other tasks that are operationally similar to the training regime, and don't transfer to a more generalised effect on intelligence. In addition, it seems that the duration of training required for even small amounts of cognitive enhancement is likely to be more than a year, rather than weeks or months of practice.[7]

In recent years, the attention of people seeking to boost intelligence has been drawn to a very different sphere. In 2008, neuroscientists acknowledged the growing demand for brain-boosting drugs, publishing a paper in *Nature*, stating that the idea of 'enhancement' should no longer be a dirty word.[8] Much of the excitement centred around modafinil, a drug usually prescribed to help with sleep disorders. In healthy people, it appears to improve aspects of intelligence, such as decision-making and learning and memory, as well as 'fluid intelligence', which is essentially our capacity to solve problems and think more creatively.

Trials of modafinil proved hugely successful, and sales of the drug soared. It's not clear exactly how it works, but it appears to have something to do with preventing nerve cells from reabsorbing dopamine once it is released into the brain. The difference is that it somehow does so without producing the addictive highs and painful crashes associated with other stimulants. It also seems to boost blood flow in regions of the brain associated with specific cognitive functions.

Of course, in the UK it remains a criminal offence to supply prescription-only medicines without a prescription, although it is not illegal to buy and import modafinil into the UK for personal use. Recent stats are hard to come by, but a 2011 poll conducted by *New Scientist* and the BBC's flagship current affairs show *Newsnight* suggested that 38 per cent of people had taken the cognitive-enhancing drug at least once.

It's human nature to want to push against our limitations, but what about the risks? The company which makes modafinil, Cephalon, insists that the drug is for treating medical sleepiness, but it's clear that modafinil is becoming a lifestyle drug for many. It does have side effects, most commonly headaches. And it certainly shouldn't be used in pregnancy as it can cause congenital malformations. Nevertheless, in 2015, researchers at Harvard University in the United States and the University of Oxford in the UK concluded that it was the world's first safe 'smart drug', after performing a comprehensive review – while acknowledging that there was limited information on the effects of long-term use.

Cognition-enhancing pills are probably here to stay. But be cautious. Their use has met with criticism from many areas. For one thing, how comfortable are you that your classmates might be using them to revise for an exam? Is it fair that your competition for your next job might use them to outshine others? And if you do choose to dabble, what happens when you stop? Which is the real you – the one who uses drugs to boost their brain power, or the one who doesn't? All these questions, and more, are worth considering if you are thinking about popping any pills to boost your brain. There are no easy answers – but we should probably start thinking the questions through, before a drug offering far more than a few percentage points of enhancement comes our way.

HOW TO UNDERSTAND MEMORY

The film *Eternal Sunshine of the Spotless Mind* made one crucial mistake. Doctors are scanning Elijah Wood's brain. They pinpoint the memory they are looking for and boom, delete it. If only it were that simple. A fully functioning memory is an essential underpinning of the sort of functional intelligence we've just been talking about. But individual memories aren't tied up in neat little packages that can be accessed or removed at will. In fact, we're not quite sure what a memory, or memory in general, looks like at all.

In the Harry Potter films, they are silver streams that can be teased from the head with the tip of a wand. In the Pixar movie *Inside Out* they are small glowing balls, stored in vast racks of shelving in our minds. But in reality memories are formed from a chaotic web of neural connections that can be strengthened and pruned depending on many different factors.

We have several different types of memory for different kinds of information. Can you remember the film I just mentioned right at the start of this section? How about what you had for dinner last night? Can you count backwards from 100 in 7s? And do you remember the name of your first pet? These are the kinds of thing you should be able to remember, either because you've only recently encountered the information or because it was something that was really important to you – like Harvey my lop-eared rabbit who was my pet when I was a kid.

Although it's difficult to say exactly, an average memory can keep around four specific things in mind at once, for up to thirty seconds.[9] Only the really meaningful stuff makes it into our long-term memory, such as a conversation that contained a personal insult or an embarrassing pratfall: we have selectively strong memories for events that are emotionally arousing. These long-term memories tend to be split into semantic memories for facts, such as the concept of a sandwich, and episodic memories about events we have experienced, such as having a seagull steal a particularly good sandwich as you were sitting on the quayside.

People who are good at remembering one kind of memory aren't always great at the others. At one extreme, some individuals are unable to form episodic memories at all. People with 'severely deficient autobiographical memory syndrome' would report an awareness of the fact that they were at dinner, but they have no feeling of re-experiencing it. It's more of a factual memory.

At the other end of the spectrum are the fifty or so known people who can recall most days of their life with exquisite detail. Ask them what they had for lunch on 7 November 1983 and they'll tell you. They'll also be able to recall what the weather was like, what they watched on TV that evening and how they were feeling about their job at the time.

Then again, you might be someone who is amazing at remembering faces – a super recogniser. At the other end of that spectrum, are people with prosopagnosia, or face blindness, who can't tell faces apart even if they're of people they know well. Most of us fall somewhere in between these extremes. And we don't always land in the same place – things as simple as stress and tiredness can severely affect our ability to remember things.

You might experience a lot of 'mind pops', random memories that spontaneously come to mind, normally in dull moments. Again, this is perfectly normal. These involuntary recollections happen to all of us, on average about twenty times a day, although there's a lot of variation between individuals. They are a basic characteristic of autobiographical memory, and once they pop into your head they soon disappear. Just like dreams, if you don't write them down, you forget all about them. Involuntary memories are often associated with the environment we are in, and there is a high probability they have relevance to what is happening at that moment. They might be an update, a nudge of sorts, reminding you of the last time you were in this situation. As we age, we tend to experience more mind pops, and retrieve fewer memories consciously, perhaps because we find it harder to inhibit thoughts as we get older.

Our memory skills (or lack thereof) are more likely to be due to the life experiences we've picked up along the way. Most of us start out with roughly the same memory ability, but subtle differences at the beginning of life get amplified by experiences and interests that build on each other. True to stereotype, women tend to acquire better episodic memories,[10] whereas men tend to remember spatial information better, and so can be better at forming mental maps of things and places.[11] Women generally perform better at verbal tasks, such as recalling word lists. Personality type seems to be a factor, too: people open to new experiences tend to have better autobiographical memory.

One thing we all have in common is that ageing affects our recall. If by the time you're forty you notice you can't remember new names, it's not that your brain is overloaded – our memory capacity is practically unlimited. Rather, gradual changes in brain structures, such as a reduction in the density of dendrites that help to form connections between neurons, make the creation and retrieval of memories less efficient. In short, when it comes to memory, all of us get a bit creakier as we age.

It's normal to worry about dementia, but there's no reason to be alarmed if you do have the odd 'senior' moment. I'll talk more about methods to stave off general mental decline later in this chapter, but the processes are perfectly normal. As well as slowing down, certain other memory skills shift to a lower gear with age – multitasking becomes more difficult, for instance. One warning sign that things aren't normal is the inability to summon a memory even when prompted. With normal ageing, it might take you longer to remember, but at the early stages of Alzheimer's having more time won't help, because the information itself has degraded. When these kinds of shifts happen, or your memory interferes with daily life, it's time to consult a doctor.

HOW TO BECOME A MEMORY CHAMPION

It sounds implausible, but study after study shows it to be true: superior powers of recall are due to well-practised strategies and memory tricks, not any innate talent for remembering. The brains of 'mnemonists', or memory champions, look like everyone else's. The techniques they use to memorise hundreds of words or digits in minutes can be learned by anyone. It is easier to become one than you think.

The strategy almost all top memorisers rely on is called the 'method of loci'. This involves imagining a route they know well, such as their commute to work, or moving around their house, and associating the information to be learned with landmarks along that route. They can retrieve the information later on by making the same journey in their mind and visualising the objects connected to each landmark.

To see if this might help all of us improve our memory, one study recruited fifty-one people who were split into three groups. One group was instructed to practise the method of loci for thirty minutes a day for six weeks. The members of the second group were told to practise holding information in their heads for short periods without being given any particular strategy to help them. The third group did no training.

At the start of the experiment, the volunteers could remember twenty-six to thirty words on average from a list of seventy-two. After six weeks, those who trained using the method of loci could typically remember a further thirty-five words. The other groups showed a small improvement – eleven extra words for the uncoached memory group and nine for the group that did no training. Happily, this had long-term benefits – four months down the line, the method-of-loci group still performed far better than the other groups.[12]

The secret of this trick is that the brain prefers storing images to words and numbers without any context – particularly if you place those images in an orderly location. The richer you can make this imagery, the more easily it will be recalled. Joshua Foer, a journalist who became the US memory champion with only a year's training, says that the techniques all come down to transforming a piece of information that lacks any sort of context, say a string of random numbers, and applying associations to it to make it meaningful. One strategy for doing that is to come up with a way of visualising whatever it is you're trying to remember and making that memory as bizarre as possible.

Sadly, while learning this method might allow you to remember a shopping list better, or impress at a family gathering, it probably won't improve your ability to function in everyday life. Many of the world's best memorisers can perform astounding feats in the memory ring, yet can be quite forgetful in everyday life, just like the rest of us.

HOW TO STOP DIGITAL MEMORY ROT

There's a huge contemporary worry when it comes to memory: the effects of technology on it. It's not a new one. Socrates famously decried the arrival of writing, saying it would erode memories and give the illusion of knowledge rather than the real thing. In the 1970s, the big worry was the use of electronic calculators in schools. Now we project our worries about the loss of our mental faculties on the digital devices most us are wedded to. After all, if the smartphone in our hand remembers all our contacts, our friend's birthdays, our appointments and how to get there, aren't we bound to lose the capability to remember things for ourselves?

Yes and no. The evidence shows that while we don't need to worry that tech is warping our memories too badly, there are a few things to be aware of. For instance, learning using screens rather than books degrades performance slightly, partly because it's tempting to browse the web, check email or watch the latest viral video. Even without distractions, we seem to get less from reading on a screen. When volunteers were asked to read a mystery story either on a Kindle or in booklet form, those who read it on paper were better at reconstructing the plot than those who read it on the Kindle: they were almost twice as good at putting fourteen plot events in the right sequence.[13]

This might be because paper books give our brain plenty of physical cues to help us remember the plot, and as we've seen memories are easier to recall when we form lots of visual associations around them. You might recall, for instance, that a sudden death occurred halfway down a left-hand page or a third of the way through the book.

We're also using digital services to outsource some of our memories on purpose. Probably the largest data dump is of snapshots of events, whether it is thousands of photos posted on social media or status updates documenting our lives. You might think that taking pictures and sharing stories help you to preserve memories of events, but the opposite is true. When researchers at Princeton University sent people out on tours, those encouraged to take pictures actually had a poorer memory of the tour at a later date. Creating a 'hard' copy of an experience, even if only in the form of a digital photo, leaves only a diminished 'soft' copy in our own heads.

The mere expectation of information being at our fingertips seems to have an effect on its retention. When we think something can be accessed later, regardless of whether we will be tested on it, we have lower rates of recall of the information itself and enhanced recall instead for where to access it.[14]

Studies like these suggest technology isn't making our memory worse, but just changing it, reflecting the fact that increasingly we don't need to remember content, but where to find it. Sometimes this might be a good thing. When people were given two lists of words and asked to memorise them in twenty seconds, those who were allowed to save the first list on a computer rather than deleting it before moving on to the next could remember more information from the second list at a later date. It seemed as if cognitive offloading freed up vital brain resources that allowed them to remember new information.[15]

But relying too heavily on devices can mess with our appreciation of how good our memory actually is. We are constantly making judgements about whether something is worth storing in mind. Will I remember this tomorrow? Does it need to be written down? Should I set a reminder? This is called meta-memory, and technology seems to screw it up. For instance, people who can access the internet to help them answer general knowledge questions, such as 'how does a zip work?', overestimate how much information they think they have remembered, as well as their knowledge of unrelated topics after the test, compared with people who answered the questions without going online. Essentially, you lose touch with what came from you and what came from the machine.

Whether any of this matters, however, really comes down to whether you have constant access to digital devices or not. If they disappear – deep in the countryside, at the beginning of an exam, in an emergency, or even a technological catastrophe, we may underestimate how much we would struggle without them.

So while we may not notice the changes to our memory, the interface between us and our devices is certainly making subtle tweaks. The brain will begin to adapt; it is so efficient that it would be silly not to use up precious real estate that's not being used for one thing, for something else. The best defence is to keep your memory skills ticking over: get out a pen and paper once in a while, leave your camera at home and take mental pictures, and revise from a textbook, not a screen.

Did I mention revision? Ah yes, because there's one situation we'll all encounter at some stage in our lives where having your mind at its sharpest can have a definite positive impact on your future. Let's look at it now.

HOW TO ACE AN EXAM

My friend, let's call her Fran, is one of the cleverest people I know. She passed all of her academic tests with top grades, has a first-class degree from the University of Oxford and was one of the UK's youngest partners in a law firm. Her ability to ace an exam is second to none. So while writing this chapter I turned to her for some anecdotal wisdom. 'What's your secret?' I asked. She laughed and wondered out loud whether it had anything to do with the shot of vodka she threw back before every exam. The really funny thing is, she might have been on to something.

But before we turn to the booze, let's look at some other techniques that might help you ace an exam. Everyone has their preferred method of revision. Some swear by colourful mind maps. Others go for flash cards. The most common practice is rereading notes and highlighting the relevant material. The million-dollar question, though – which provides the biggest pay-off from those hours of hard graft?

Actually, science suggests that one technique stands head and shoulders above the rest when it comes to memorising things for a specific, short-term purpose: simple recall. Although it is more than two millennia since Aristotle wrote that 'repeatedly recalling a thing strengthens the memory', cognitive scientists have only recently come to appreciate the effectiveness of this so-called retrieval practice.

In a landmark study from 2008, researchers from Indiana asked forty students to learn the meaning of forty Swahili words. Despite receiving no feedback on whether they were correct or not, those who were repeatedly asked to recall the words during the learning session aced the final test a week later, with an average score of 80 per cent. In contrast, those who repeatedly studied the words, without actively testing themselves, scored a measly average of 36 per cent.[16]

Later experiments showed that retrieval practice also outstrips more active methods of study, such as drawing weird and wonderful bubble diagrams to represent information in a passage of text. And it's not just useful for language students – schoolchildren, medical students and neurological patients in cognitive rehabilitation all do better at tasks when they test their memory at regular intervals.

It also matters when you study. One of the easiest ways to increase how much you remember without any additional effort is to carefully time when you sit down to do it. We've already spoken extensively about the benefits of power-napping in Chapter 7, but numerous experiments have found that sleep shortly after learning new facts or skills helps the brain reinforce its memory of recent events – whether that sleep is a good night's heavy slumber or, yes, just a well-timed afternoon nap.

You might also want to consider the timing between different study sessions. We learn much better if we revisit material after an interval rather than hammering it home during a single session. Surprisingly, the length of this interval determines how much you remember – get it right, and it could easily bump you up a grade or two. The sweet spot seems to be revisiting your subject at a time equivalent to between 10 and 20 per cent of the interval between learning the material and subsequently taking the exam. So if you are revising for a test in twenty-four hours' time, for example, you should revisit the material roughly two to five hours after you first reviewed it. Do this, and studies show that exactly the same amount of revision time could give you at least a 10 per cent higher test score.[17]

Finally, was Fran right? Could a little drop of the hard stuff really help? Well, it depends on your goal. A small tipple, the equivalent of a single vodka and coke, does indeed make it easier to solve problems that require creative solutions, or thinking outside of the box. The theory is that alcohol reduces the capacity of your working memory – in other words, your ability to focus on one thing while blocking out peripheral information. By lowering these walls, your mind can wander, making novel connections that would otherwise be overlooked – what's thought to be the basis of creativity in the brain.

However, a perhaps obvious word of caution: as is well-known, alcohol can also impair executive control, which is necessary for tasks that involve detailed planning or decision-making, or to answer questions which are technically difficult or require you to overcome an instinctive response. And let's face it, swigging from a bottle of vodka outside the entrance hall or interview room probably isn't your best look.

HOW TO STAVE OFF MENTAL DECLINE

It may not be something you need to fix right now, but we all want to keep our mind sharp as we get older. It's never too early to start the good habits that will help us in our golden years. The brain is often likened to a muscle, for good reason: give it a good workout and it stays strong. But as I mentioned earlier in the chapter, brain-training apps have largely been debunked; instead the trick to improving our cognitive fitness is to find activities that boost what's called our 'cognitive reserve'.

You can think of this as some mental padding, spare neural capacity that allows your brain to sustain more injury before you feel the effects. It is why two people with the same amount of damage to their brain can see unequal impacts on their daily life. Cognitive reserve has been linked to a higher IQ and greater education. This may be why rates of dementia have fallen in developed countries over the years (even if, with more older people around, absolute numbers have been going up).

One option that appears to be beneficial is to be socially active. Some evidence suggests that being married is strongly associated with a reduced risk of cognitive decline and dementia, in theory because of the regular conversation and mental effort involved in maintaining a good relationship.[18] But there's not one type of social contact that's better than another – just another reason to work on your friendships whenever possible (more on that in Chapter 4).

Something else you might want to try is learning a language or playing a musical instrument. People who are bilingual develop dementia later than monolinguals and musical training seems to protect some areas of the brain from decline later in life. A sense of purpose seems to be helpful in staving off mental decline, too. In one of the largest studies looking into this, 3,500 people were asked about their sense of purpose in life. Those whose answers suggested they felt that their life had direction and their actions were aided by having goals tend to perform better when given memory and cognitive tasks. It might be that a sense of purpose provides motivation and pushes people into activities that help build cognitive reserve. It also tends to make you more active and more sociable.

We've already seen in Chapter 6 how exercising your heart, muscles and lungs can boost brain chemicals that help ward off dementia, how a good diet can add years of healthy cognitive function and how sleep can help clear out potentially damaging brain gunk at night. But perhaps one of the most effective methods of staving off mental decline is completely left field. The science is new, and not completely settled yet. Even so, you might want to give some consideration to how well you are brushing your teeth.

In 2017, Bei Wu at New York University published details of a strange study. She had followed the lives of 8,000 people in China for thirteen years, recording their cognitive function and tooth count. She found a strong correlation between tooth loss and a drop in cognitive function, even after accounting for the natural changes that occur in both with age. It didn't get a huge amount of attention, but perhaps it should have. Fast forward a few years and a landmark paper was published offering compelling evidence that Alzheimer's may be caused by a bacterium involved in gum disease.

For decades the accumulation of two types of proteins in the brain – amyloid and tau – has been the focus of researchers studying the disease. The proteins form sticky plaques and tangles that destroy neurons. But it has become obvious that trying to clear these proteins isn't working. Alzheimer's drug development has had a 99 per cent failure rate. The lack of results has been compounded by the discovery that people – including some in their nineties with exceptional memories – can have plaques and tangles without dementia.

Then it turned out that amyloid functions as a sticky defence against bacteria. When you inject bacteria into the brains of mice engineered to make Alzheimer's proteins, plaques develop

THIS BOOK COULD FIX YOUR LIFE

around bacterial cells overnight. Gum disease was already known to correlate with Alzheimer's, but was thought more likely to be a symptom. This insight, however, shone a spotlight on the main bacterium involved in gum disease called *Porphyromonas gingivalis*. Studies have shown that this bacterium invades and inflames brain regions affected by Alzheimer's; that gum infections can worsen symptoms in mice with Alzheimer's; and this can cause Alzheimer's-like brain inflammation, neural damage and amyloid plaques in healthy mice.

In 2019, researchers reported finding the two enzymes that *P. gingivalis* uses to feed on human tissue in 99 and 96 per cent of fifty-four human Alzheimer's brain samples taken from the hippocampus – a brain area important for memory. These protein-degrading enzymes are called gingipains, and they were found in higher levels in brain tissue that also had more tau fragments and thus more cognitive decline.

If that wasn't enough, when the researchers looked for signs of *P. gingivalis* in the brains of healthy people, they found some, but at low levels. This supports the theory that *P. gingivalis* doesn't get into the brain as a result of Alzheimer's – but might be the cause.

As a well-informed reader, you'll know that caution is still required in saying that correlation implies causation: not every link between two factors means the one causes the other, and both could have an entirely separate cause. Our history of failed theories and drugs associated with Alzheimer's is perhaps a further reason to stay cautious. But in this case several experiments are converging to suggest that gum disease really might be behind the disease. For instance, when the team gave *P. gingivalis* to mice, the animals developed dementia. And when the team gave these mice a drug that kills the bacteria, there was a reduction in the symptoms.

There are hints that the same may be true in humans: the more gingipains a person has the more symptoms of Alzheimer's they have. It is not necessarily the only cause, of course, but for now it might be wise to prevent gum disease just in case. At the very least, the side effect of brushing with an electric toothbrush, flossing and regular trips to the dentist is a brighter smile.

Of course, one day we might be able to physically replace lost brain cells with a transplant of fresh neurons – human trials are underway to treat Parkinson's disease using foetal brain cells with the aim of growing them into replacement neurons. Or we might be able to zap our brains with electrical stimulation to improve specific cognitive functions such as memory, creativity or mathematical ability. Or perhaps we'll all be taking an infusion of blood from young people, a process that has shown some promise in experiments with animals and is now being investigated as a treatment for Alzheimer's disease. But I wouldn't count on any of these things being available over the next few decades. Spending more time with friends, brushing your teeth and working on your lust for life seem a lot more pleasurable in the meantime.

TOP TIPS FOR GETTING SMARTER

🔧 Stay in education longer – each additional year can boost IQ by between 1 and 5 points.

🔧 The brain prefers to create memories with images, rather than words, so next time you need to remember something important, link it to memorable or funny images rather than concepts or facts. Need to remember to pick up some eggs? Think about cracking one over your head, rather than a verbal shopping list. You'll find it easier to recall once you're at the shops.

🔧 Keep your old-school memory skills ticking over by eschewing digital aids every once in a while: get out a pen and paper, leave your camera at home and take mental pictures, and revise from a textbook, not a screen.

🔧 If you're revising for an exam, revisit the information at a time equivalent to between 10 and 20 per cent of the interval between learning the material and subsequently taking the exam – it could give you at least a 10 per cent higher test score.

🔧 Be socially active: the regular conversation and mental effort involved in maintaining good relationships can help keep your brain in trim.

10

HOW TO BE WISER

I SAID AT the beginning of the previous chapter that this one would follow on naturally from it. It acts as something of a counterpoint, as we examine those elements of intelligent behaviour that go beyond, well, raw intelligence as it might be measured in an IQ test.

So, in the spirit of what follows, it's only right that I start this chapter with a story that acts as a counter to the tales of my childhood precocity with which I started the last one. It was an incident that weighed on me so heavily for years afterwards. One evening when I was in my early teens, and my parents were out, the telephone rang. The man on the other end of the landline was friendly. He was from our insurance company, and asked whether I had time to confirm a few things about the house for our policy. I told him that my parents weren't in. He said that was OK, and that I probably knew the answers myself.

I set about trying to help. It started with the number of people in the house, the number of rooms and how many cars we had. He asked about alarms, and locks on the doors. It was only twenty minutes later, when we had got on to specifics about exactly what kind of jewellery was kept in what rooms, that it dawned on me what was happening. I panicked. I slammed the phone down, my heart beating nineteen to the dozen, and never mentioned it to my mum and dad.

I was only young, yes, but I was also supposedly so super-bright. How could I have been so easily duped? The answer is perhaps obvious – the sort of intelligence that get you top marks in tests is only one part of the picture. Simply having a high IQ doesn't make anyone immune to making silly mistakes.

It's a fact that anyone who considers themselves intelligent will have to bear in mind: humans are terrible at thinking. We all make bad decisions and sometimes terrible errors in judgement. We are hopelessly biased and beholden to certain set ways of thinking. We consistently overrate our own abilities, while sometimes lacking the cognitive tools to empathise or analyse situations clearly.

Being exposed to a lot of the science surrounding our flawed ways of thinking has made me revisit the question of how intelligent I really am. I'd encourage anyone to do the same. Beyond the more measurable elements of intelligence and memory recall, a little bit more of what we might term 'wisdom' can help us in all sorts of areas of our lives, from academic exams to our social and moral obligations, from choosing a new car to avoiding being conned.

HOW TO UNDERSTAND WISDOM

After my experience with the dodgy insurance guy, I should probably have some sympathy for Paul Frampton. A 68-year-old divorcee, he was delighted to strike up a friendship on an online dating site with someone claiming to be the Czech glamour model Denise Milani. She was beautiful and seemed head over heels for him. They arranged to meet during one of her modelling assignments in South America. However, when he arrived in La Paz, Bolivia, he was disappointed to find that Milani had been asked to fly to another shoot. But could he pick up the suitcase she had left? He did, and was arrested on arrival in

Buenos Aires, Argentina, and charged with smuggling two kilo-grams of cocaine.

It may seem like an obvious honeytrap, yet Frampton wasn't exactly lacking in brainpower. An acclaimed physicist, he had written papers on string theory and quantum field theory. How could someone so intelligent have been so stupid? Psychological research shows that Frampton's behaviour isn't as exceptional as it first appears. As we've seen, IQ does correlate with many important outcomes in life, but intelligence and expertise can sometimes make us more likely to make stupid errors.

Wisdom, like intelligence, is difficult to define, but broadly it amounts to the ability to avoid making those stupid errors: it is a suite of mental faculties that allows us to avoid blunders in thinking and take 'good' decisions. It implies an ability for critical thinking and the capacity to assess risk and uncertainty and weigh up conflicting evidence; it integrates knowledge and experience with self-knowledge and an ability to understand the motivations of others. It's almost akin to being able to take an outside view of your own intelligence – not just of its strengths, but of its limitations; knowing not just what you know, but what you don't know; an awareness of the biases to which you and others are susceptible that cloud good judgement. In other words a lot of elements are involved. And while psychol-ogists agree that wisdom is generally acquired through experience, that alone is not enough: experience does not auto-matically make you wise.

As to what leads to incongruities between intelligence and wisdom, there are many suggested explanations. One that's gained popularity in recent years is that people who make poor deci-sions rely too heavily on intuitive, 'system-1', rather than deliberative, 'system-2' thinking. Simply, they may have the capacity for abstract reasoning, but they aren't engaging it to think through life's problems. I'll talk more about the difference between these two modes of thinking later in this chapter, as well as some of the flaws associated with this binary model of thinking.

Luckily, science can help us to sidestep some of the biggest traps that reduce our capacity for wise thinking. That starts with recognising some common flaws in the way humans think.

HOW TO AVOID COGNITIVE BIASES

So, we want to look at some of the ways human thinking is commonly biased – divorced from reality. So try this one out for starters: do you think you are less likely to make a biased decision than the average person?

While you're considering that question, think about which product you'd prefer – one that said it was '95 per cent fat free' or one that advertised being '5 per cent fat'. If you went for the first one, like most people, you've been duped. This is a common trap known as a framing bias. It takes advantage of our tendency to view certain statements and statistics more favourably depending on the way they are phrased. (This is, incidentally, also a trick you can use to your advantage if you want to persuade people of the strength of your arguments – see the section on persuasion in Chapter 11 for more on that.)

Then there's the sunk-cost bias: the tendency to pour more resources into a failing project so as not to sacrifice your initial investment, even though it will ultimately cost you a lot more than simply giving up. Or there's the gambler's fallacy, the belief that chance events somehow even themselves out. If the roulette wheel lands on a red three times in a row, then how many of you would feel like your next bet should be placed on black?

Oh, and if you came to the conclusion you were less likely to fall prey to biased thinking, that's the illusory superiority bias, or the 'above average' effect. This is the belief, harboured by many in the West at least, that they are above average at everything from driving to performance at work. In one classic study, 93 per cent of drivers in the United States rated themselves as above average on their driving ability. In fact, the effect first came to light among a small sample of drivers in an earlier study who all rated themselves as above-average – despite having all been recently hospitalised in car crashes.

The net effect of this and many other similar biases is that, even though we believe ourselves to be so intelligent, we're often very daft. That phenomenon even has its own name – dysrationalia. One root cause of this phenomenon might be down to people lacking the skills to regulate their emotions and allow them to dissect and account for their feelings: to escape the sunk-cost bias, for instance, you need to overcome your regret at the loss of your earlier investment.

There's a good body of evidence to suggest that one good way to overcome our biases is to mistrust our own expertise. Some studies suggest that the brightest students, as measured by exam scores, tend to have larger bias blind spots – they underestimate their own capacity for bias compared with others.

Meanwhile Nate Kornell at Williams College in Massachusetts presented a group of mathematicians, historians and athletes with certain names of significant figures within each discipline and asked if they knew who they were. The participants were far more likely to claim recognition of the figures supposed to come from their own discipline – even if they were fake.

Matthew Fisher, then at Yale University, saw something similar when he quizzed university graduates on the subject in which they majored. He wanted to check their knowledge of the core topics of the degree, so he first asked them to estimate how well they understood some of the fundamental principles of their discipline. A physicist might have been asked to gauge their understanding of thermodynamics, for instance, or a biologist to describe the Krebs cycle.

Fisher then sprung a surprise test. The participants had to write a detailed description of the principles they claimed to know. Despite their claimed knowledge, many stumbled and struggled to write a coherent explanation. But when they considered topics beyond their specialism, or more general, everyday subjects, their estimates of their knowledge tended to be more realistic.

This is worrying, because further research has shown that perceptions of expertise can lead to what's known as 'earned dogmatism' – the sense that you have earned the right to remain closed-minded about a subject, while rejecting arguments that disagree with those views. A politician, say, may have outdated theories based on their degree in economics, but thanks to their earned dogmatism they may ignore new information.

Earned dogmatism can also be seen in 'Nobel disease': the tendency for Nobel Prize-winners to develop bizarre theories later in life. A notorious example was Kary Mullis, who pioneered the polymerase chain reaction that is now essential for genetics research, but later came out as a climate change sceptic and AIDS denialist. Or there was James Watson, one of the discoverers of DNA, who still holds outdated and prejudiced views about racial differences in intelligence, despite frequent criticism from respected colleagues. The undoubted status that comes from winning a Nobel Prize allows them to deny even the most basic evidence that runs contrary to their opinions.

Another area where your wisdom may falter is in your judgement of your own affairs compared to others. This is called Solomon's paradox, from King Solomon, who ruled Israel some 3,000 years ago and was renowned for his wise judgement of other people's dilemmas. Stories suggest that he was less astute when it came to his own life: he kept hundreds of pagan wives and concubines, against the instructions of his religion, and he failed to educate his only son, who grew up to be an incompetent tyrant who ultimately contributed to the kingdom's downfall.

It seems that we are all fallible to this problem. Just like Solomon, we tend to find it easier to reason wisely about other people's dilemmas than our own, meaning that our thinking is often worst when it may matter to us most. A pioneering test of this, created by Igor Grossmann, a psychologist at the University of Waterloo, Canada, presents people with various dilemmas, both political and personal. They are then scored according to criteria such as willingness to explore other conflicting points of view and capacity to recognise the inherent uncertainty in a situation, rather than thinking in dogmatic, absolute terms. Intellectual humility is another criterion: whether you admit your ignorance and show a desire to find out more information. Grossmann discovered that overall scores are only loosely linked to conventional measures of intelligence, and that 'wise-reasoning' scores tend to be better at predicting overall health and well-being than IQ.[1]

Once you give this inability to apply wisdom in our own affairs some thought, you'll probably notice it cropping up all over the place. While I was writing this chapter, I spent a great deal of time dishing out advice to my friends on how to balance work and home-schooling, while we were in the midst of the pandemic lockdown. It was easy to tell others not to worry, that the science shows that games, activities and cartoons were all helping their toddler-age children learn new words, colours and concepts. Trying to apply that same logic to my own daily schedule, however, was a constant challenge.

Thankfully, there are a few simple steps we can take to think more wisely and with less bias. First, when thinking about problems, from political to personal, imagine that you are discussing someone else's life rather than your own. This self-distancing restores a less biased, more open-minded attitude.

Next, repeatedly question your judgement over a situation. All that mindfulness practice you've put in after earlier chapters should come in helpful now, too – it seems to encourage a wiser, more rational stance, reducing errors such as the sunk-cost fallacy.[2]

Lastly, there is good evidence that with practice you can learn to identify your own biases and logical fallacies, so you can spot the potential for dysrationalia before you go wrong. One of the most robust tricks is to pause to consider the opposite of what you had just been thinking, actively challenging your assumptions and intuitions and looking for alternative hypotheses.

You can do this yourself, but you might want to teach it to your children and encourage it in your workplace too. You and your colleagues may think you're gifted, or surrounded by intelligent people, but unless you make some serious mental effort, you're each just as fallible as the next person. You've been warned.

HOW TO KNOW THE REAL YOU

Implicit in what we've just discussed about recognising cognitive biases is capacity for self-knowledge. A lot of self-help relies on you being able to make accurate judgements of yourself – yet this is incredibly difficult to do. It turns out that the flaws in our thinking start with trying to evaluate ourselves.

Take the illusory superiority bias, which we encountered in the previous section. Research into the phenomenon indicates that we are particularly deluded when it comes to judging the traits we care about most – those we consider highly desirable or highly undesirable.[3] For instance, we'll see later how students' self-reports of intelligence bear almost no relation to their results from IQ tests. Those with high self-esteem overestimate their intelligence (I was definitely top of the class as a kid, honest!), while those with low self-esteem underestimate theirs.

This suggests that when it comes to the qualities we associate most with self-worth, such as intelligence but also physical attractiveness, our judgements can be based more on confidence than objectivity. (By the way, if it's confidence you lack and you want to build it, I'd suggest a reread of Chapter 3.) On the surface, this tendency to think well of ourselves seems like a good thing. And the evidence shows this mental strategy does have its advantages. If we think we are better than average, it makes us feel good, which in turn protects our mental health. Being overconfident can also enhance your social status.[4] On the flip side, underestimating your abilities might protect you against failure if it makes you less likely to test yourself.

But there are reasons to be cautious. Any benefits pale when set against the costs of self-misperception. To advance in life and society, we have to make choices about where to invest our efforts, and on what outcomes to stake our self-esteem. Inaccurate self-knowledge leads us to make poor choices, contributes to conflict with others, and ultimately causes us to fail in our endeavours.

So how can you tell whether you are in thrall to a superiority complex, and what can you do to avoid it? There are a few pointers. If you are judging a trait in which you have expertise or experience, then your self-knowledge is probably good. For example, doctors more accurately judge the extent of their medical knowledge than lay people. Likewise, professional athletes can better perceive their sporting skills than amateurs. Conversely, the young and inexperienced are more likely to demonstrate poor insight in areas of life in which they have yet to gain experience.

Of course, it might be that you are indeed smarter than the average human. But when you're considering what you're good at, it's probably best to err on the humble side, unless you're talking about your specialist subject – or take a test to figure it out, rather than trust your own judgement. But it's important also to remember that it's not just about who you are, but how others see you. The people you interact with hold up a mirror in which you see yourself – so, how clearly do we perceive that reflection?

The good news is that most of us have reasonable insight into how people see us. The bad news is that we're not great at knowing what a particular individual thinks. One reason for this is that we assume others know more about us than they do. Studies suggest that, for example, if you ask someone to consider how strangers would judge your skill at playing darts, you'll tend to come to a conclusion using memories of past performances that strangers know nothing about. You have difficulty knowing how someone views you because you know too much about yourself.

It gets stranger. One long-running study, known as the Concordia Longitudinal Risk Project, asked children between the ages of six and fourteen to evaluate themselves and their classmates on scales of aggression, likeability and social withdrawal. Two decades on, the ratings from peers were much more closely associated with the personality attributes of people as adults than the self-generated ratings.

When we look at this in more detail, it seems that being immersed 24/7 in your thoughts and feelings may give you greater insight into private attributes, such as neuroticism and conscientiousness, but it's a hindrance when it comes to rating aspects of your personality that are relatively easily observed. For instance, if you ask me how kind and considerate I am, I'm likely to incorporate how kind and considerate I intend to be with how considerate I feel inside, rather than focusing on how kindly and considerately I actually act.

Others make their assessments judged on your actions, so some of their assessments are more accurate than our own. This is particularly true when it comes to our skills and abilities. We only have moderate self-knowledge in areas such as academic ability, vocational skills and sporting prowess.

However, studies reveal one exception: foreign language ability. Here we have superior self-knowledge – probably because of the constant feedback we get from talking in a second language. I am certain, for instance, that my French is absolutely useless, from the frequent quizzical looks and insistence of French-speakers on talking English back to me. Unfortunately, we don't tend to get a lot of useful feedback like this in other areas of our life. Feedback might be forthcoming from our bosses but outside of work it's not regularly offered.

The answer, then, is to actively seek it out if you want to

benefit from the wisdom of others to improve your self-knowledge. But remember, if you're going to do this, that how others see you also depends on what they are like themselves. People who are well adjusted, for instance – who positively view various aspects of their lives, themselves and their relationships – are more accurate in their assessments of others than people who are less well adjusted.

Studies suggest that simply asking people to try to form more accurate impressions is actually the simple answer to all this, presumably because it increases how much attention you're paying to others, rather than relying on your – sometimes untrustworthy – gut instinct.

And one final piece of advice: the people closest to us may have a lot of information, but a few of them are also the most biased. This is especially true of your parents: when it comes to their opinion of you, take anything they say with a large pinch of salt.

HOW TO AVOID UNCONSCIOUS BIAS

The death of George Floyd under the knee of a police officer in Minneapolis in May 2020 shook the world to attention and gave a new impetus to the Black Lives Matter movement, but it was no isolated incident. Every day there are stories of people being treated with suspicion – or far worse – based on their skin colour while going about their daily lives.

This is in spite of the fact that, for the past 40 years, opinion polls show a steady decline in racist views in the United States, the UK and other countries. This has led some researchers to suspect that, as explicit racism has been driven underground, unconscious or 'implicit' bias is playing a crucial role.

In psychological research, the label 'implicit' refers to processes that aren't direct, deliberate or intentional self-assessments. When we can't retrieve a memory explicitly, we might still behave in a way that is shaped by our past experiences, for instance, or by deep-rooted cultural stereotypes. The conscious mind governs deliberate actions, rational thoughts and active learning, while the unconscious carries on with processes that occur automatically or aren't available to introspection. The unconscious is a busy place: the brain is capable of processing approximately 11 million bits of information every second, but our conscious mind can handle only 40 to 50 of them.

As all of this information comes in, our brains categorise it without our deliberate attention. When we process information on a more superficial level – when we are in a hurry, tired or distracted, for example – we are more likely to rely on existing templates. Occasionally, such cognitive shortcuts can be useful, such as when we need to decide something quickly. But they can also be problematic, especially if these shortcuts were formed based on mistakes, misinterpretations, stereotypes or other biased information. When we use them, we may then be relying on and reinforcing these very mistakes and biases. When that happens with people in positions of power and authority, it can have far-reaching consequences, from discriminatory hiring practices to poorer healthcare treatment or prejudice in the legal system.[5]

We are still getting to grips with the most effective ways to identify and address bias. What is clear is that it is a difficult task that requires concerted, consistent effort. But there are strategies that make a difference.

A first step is to make biases visible. This can include using tools such as the Harvard Implicit Association Test to measure the strengths of implicit links our brains make between different concepts and words. For instance, in a racial bias test, participants would be shown black or white faces and asked to pair them with descriptors such as angry, clever, good and bad.

It's worth bearing in mind, however, that the Implicit Association Test has come under criticism for a lack of reproducibility, and the way it is often administered as a one-off tool to 'diagnose' prejudice, something it was not originally tended to do. The use of such tools needs to be complemented by active reflection – including recognising triggers for bias and examining how our life experiences have shaped our biases.

Research has shown that using blind or anonymised hiring practices may help weaken biases that can limit opportunities for women and minority groups. One study found that using blind auditions increased the likelihood that female musicians were hired by an orchestra by up to 46 per cent.[6] Research in France, Germany, Sweden and the Netherlands has showed that removing names from applications increases the likelihood that candidates from minority groups will be invited to interview.[7]

We can tackle generalised assumptions by being clear that a particular attribute is associated with an individual rather than their whole group; for example 'This boy is good at maths'. This approach can help to diminish stereotypes and the pressure to conform to them.

Taking our time with important decisions can also help us avoid cognitive shortcuts that perpetuate bias. When this isn't possible, research with police has shown that rehearsing reactions to high-stress situations can help prevent biased snap decisions.

Finding ways to identify with members of different groups by forging links with your own sense of self can diminish bias. In one study, nurses from diverse ethnicities who were shown videos of white or black patients in pain recommended the same amount of pain relief regardless of the patient's race if first asked to imagine how they felt.[8] When not prompted this way, the nurses suggested more pain relief for white patients. Metaphorically stepping into someone else's shoes can have a big impact.

HOW TO THINK FAST AND SLOW

Here's another question to test your mental acuity: a bat and a ball together cost £1.10. The bat costs £1 more than the ball. How much does the ball cost? Quick, now!

If you instinctively said 10 pence, you're in smart company: more than half of students at Harvard University and MIT jumped to the same conclusion. But think about the problem more carefully, and you realise the answer is actually 5 pence.

For years, this puzzle has been held up as the perfect example of the way we are ruled by two types of thinking: fast and intuitive, versus slow and analytical. If you arrived at the wrong answer before you had time to really ponder the problem, you might blame it on intuitive thinking leading you to make a snap judgement before slower, rational thinking had kicked in. It's the same sort of process that is sometimes blamed for our prejudicial and biased thinking, as we have just seen.

This idea that our thoughts can be split into two distinctive camps has become so popular it now influences many areas of everyday life. Marketing types try to tap into our automatic impulses with emotive adverts and special offers, while governments attempt to appeal to our deliberative sides by doing things like forcing calorie counts to be displayed on menus.

These 'nudges' are often based on the assumption that fast, intuitive thinking is likely to get you into trouble, so we need to cultivate the slower kind. The US National Academies of Sciences, Engineering, and Medicine and the World Bank have both issued reports urging decision-makers to use the slower type of thinking to avoid the expensive, or deadly, consequences of fast ways of thinking.

But books and advice that focus too much on fast and slow thinking might need a rethink themselves. A more complex picture of our mental processes has started to emerge, and categorising all our thoughts as one of these two types might in fact be leading us astray on all sorts of policies and practices.

But let's start at the beginning. It was Nobel Prize-winner Daniel Kahneman who turned the idea of us having two opposing factions of instinct and conscious reasoning into a popular concept in his 2011 bestseller *Thinking, Fast and Slow*. In it, he describes our mental processes as typically belonging to system 1 or system 2. You can characterise the difference as what distinguishes a joke and a riddle. A good joke is funny without needing to think about exactly why. This taps into system 1. A good riddle, however, requires us to lean on system 2, taking some time and brow-furrowing to get to that moment of satisfaction.

Since the book's publication, the dual-process model of the mind has blossomed into one of the most widely accepted ideas in psychology, used as a framework to investigate the source of our behaviours and beliefs. Research on binge drinking, for example, suggests that this impulsive behaviour is caused by an overactive system 1. It has also been identified as the source of interviewers' false impressions of job applicants. The model has also been at the heart of evidence given to policy-makers about the implicit effects of advertising, for example, online and on social media.

Evidence that 'system 1 bad, system 2 good' was an oversimplification comes from several directions – for example, in medicine. For several years, there was consensus that diagnostic errors were primarily caused by system-1 reasoning, and clinicians were advised to think more slowly. However, more recent reviews have found that experts are just as likely to make errors when attempting to be systematic and analytical.

And pigeonholing thought processes might have unwanted negative effects too. Believing that your decision-making process is purely down to one of two systems can lead to what's known as the 'Prince Charming' effect, in which we separate automatic thoughts from conscious, intentional ones to absolve – or rescue – ourselves from our mistakes and biases. Take the bat-and-ball problem. People tend to feel better if they believe that coming up with the answer of 10 pence was an uncontrollable, unconscious response.

At a personal level, this might seem a minor consideration, but it raises serious questions for society as a whole if we accept that uncontrollable responses excuse bad behaviour or poor decision-making. If someone commits a crime in court and says they didn't do it intentionally, it was an automatic response, a jury may recognise that and be inclined to see them as less responsible for their actions.

And there is some evidence that the idea that system-1 thought processes exist that can't be consciously accessed is itself a little wrong-headed. Take the Harvard Implicit Association Test, the test we mentioned in the previous section that is widely used to reveal unconscious biases. It involves participants matching two target concepts as quickly as possible. They typically react faster when pairings are more strongly associated in their minds, reflecting implicit attitudes and stereotypes. In 2014, however, participants were asked to predict the strength of their biases before taking the test. It turned out their guesses were pretty accurate. Our biases aren't as unconscious as we would like to believe.

Findings like these have led some scientists to question the binary model of our minds more generally. There are plenty of examples of decision-making that don't neatly fit into the two categories. Take language. We deliberately communicate, but in the flow of conversation we don't consciously rehearse what we are going to say or the grammatical rules we need to use. It is intentional and, at the same time, unconscious. The same can be said of driving on a familiar route, typing, or playing a well-rehearsed tune on an instrument.

Other studies are further blurring the boundary between conscious and unconscious mental processes. In an experiment in which participants were given an identical-tasting drink containing either glucose or a calorie-free sweetener, those who had consumed sugar perceived a hill to be less steep when asked to estimate its slant. It indicates that on an unconscious level, your body is telling you how the world looks based on what it is capable of at that moment. Unconscious factors can still influence our perception even when we give a question a lot of thought.

At the very least, we perhaps need to dispel the 'good/bad fallacy' that we tend to associate with fast and slow thinking. The assumption that because system 1 is automatic and unconscious, it is error-prone, whereas system 2 is analytic and therefore correct. System 1 can often function beautifully – it allows us to perform well-practised skills more effortlessly, for instance. You only need to start thinking about your golf swing, your tennis serve or your best dance move to quickly realise you often perform worse when you give things more thought.

Often, we end up 'overthinking'. For instance, in one study

looking at deliberation, four cars were described using positive and negative attributes. The terms were 75 per cent positive for the first car, 50 per cent positive for the next two and just 25 per cent positive for the last vehicle. After reading the descriptions, some people were told to think about the cars for four minutes before choosing their favourite, while others were asked to solve anagrams during that time. When the list of attributes was long – twelve rather than four – the anagram group of 'unconscious thinkers' consistently made better decisions than those who pondered the information. Similar tests have shown the same with assessments of things as varied as job applications and even strawberry jam, implying it might often be beneficial to delegate such complex matters to the unconscious. Go with your gut, as it were.

But if we can't always trust our instincts, but our conscious mind isn't all it's cracked up to be either, how should we think about how we think? Actually, it's quite simple: don't consign any type of reasoning to the scrapheap, but be more critical about your thinking in general, whether instinctive or deliberative. Look at how well you've scrutinised the information available, to what degree you're motivated in making a specific decision and what your motivations are. Just saying 'Why do I think that?' or imaging yourself in the opposite position can help you make a better decision. Or if it's appropriate and you have the time, invite other people to say what they would do.

Of course, for most day-to-day choices, such measured deliberation isn't always possible. So if you fell for the bat-and-ball problem, don't beat yourself up about getting it wrong – you'll know better next time.

HOW TO BE EMOTIONALLY INTELLIGENT

Riley is moodily picking at her dinner. Noticing that something is amiss, her dad asks how school was. Inside Riley's brain, a small green girl called Disgust flicks a switch, and Riley rolls her eyes: 'School was great, all right?' she replies sarcastically. Sitting at the control panel in Dad's mind, a skinny man called Fear reports the eye-roll to a character named Anger, who seems to be in charge. 'Make a show of force,' he orders. 'Riley, I do not like this new attitude,' Dad responds. The situation escalates until Riley screams: 'Just shut up!' A big red button inside Dad's head is pressed: 'That's it. Go to your room!'

I often think about this brain's-eye view of emotions in Pixar's movie *Inside Out* when I get annoyed at my toddler for throwing her dinner plate across the floor and bursting into a tantrum. It often feels like our emotions control us, that they are powerful, primal forces we struggle to understand in ourselves and others. Popular though this idea may be, it is one that psychologists would like to dispel. Other animals may be slaves to emotion, but human emotional life is more complex and cerebral, they argue. What's more, mastery of your emotions is important not just for psychological well-being, but also for success in many areas of life. The good news is that if you think you're an emotional dunce, there are ways to improve.

The concept of 'emotional intelligence' surfaced in the late 1980s, when two American psychologists, Peter Salovey of Yale University and John Mayer of the University of New Hampshire, were casting around for a pithy way to sum up human qualities such as empathy, self-awareness and emotional control. The phrase they hit upon – emotional intelligence – languished until writer

Daniel Goleman nailed it to the mast of his 1996 bestseller *Emotional Intelligence: Why it Can Matter More Than IQ.*

The concept was an instant hit: tantalising us with the idea that we each have an EQ to our IQ, and promising to let us measure how emotionally clued-up we are. (The idea of emotional intelligence is actually something Aristotle pointed out thousands of years ago: 'Anyone can be angry – that is easy. But to be angry with the right person, to the right degree, at the right time, for the right purpose, and in the right way – that is not easy,' he wrote.)

People rushed to find out where they were on the EQ spectrum. Even today, you'll find hundreds of tests online that supposedly let you test your emotional wisdom. However, the idea immediately hit problems, not least in suggesting that people with a low emotional quotient are saddled with it. Quite often, EQ tests also failed to do what they promised: allow employers to find the most emotionally savvy candidate for the job. As a result, psychologists fell back out of love with emotional intelligence. But they didn't leave us hanging. Instead, they identified three skills that can help us all become more emotionally adept, and reap the benefits.

The first skill is around perception, and is the bedrock on which the two other skills rest. Perceiving the complex range of emotions that humans can produce is not as straightforward as it might sound. Early tests of this skill used pictures of faces showing different emotions, and indeed many of the online tests you see today still use this basic way of working out how good you are at perceiving emotion. However, they fail in providing an accurate picture of your emotional intelligence. The expression of emotion extends beyond the face to gestures, movements and tone of voice and the way all these things interact.

Back in the 2010s, Katja Schlegel, then a PhD student at the University of Geneva in Switzerland, developed what could be a better way of assessing how we judge emotional cues in everyday life. Named the Geneva Emotion Recognition Test (GERT), it involves a series of short videos of actors expressing an emotion by uttering meaningless syllables. People's scores can range from 0 to 1, and preliminary research suggests that they are meaningful. When Schlegel invited pairs of strangers to negotiate a work contract, those with higher scores both negotiated more successfully and were perceived as being nicer and more cooperative than people with lower scores. Emotion recognition seems to be an important skill because it is difficult to convince a person of your ideas if you're not paying attention to their needs and interests.

Of course, it's a skill that will come in handy in all social interactions. And thankfully it is possible to improve your perception skills by practising on video clips that teach people to look for appropriate cues in the face, voice and body before getting feedback on their performance. One study found that undergraduates trained in this way achieved an average GERT score of 0.75, compared with 0.6 for controls.

Music training, too, may help: adult musicians are better than non-musicians at judging emotion in someone's tone of voice. Brain-imaging studies suggest this reflects more than a general sensitivity to basic aspects of sound. But it seems that for all of us, training to 'hear' emotion from the way music is played can modulate brain responses known to be more associated with emotions and with our ability to interpret others' minds.

Perceiving emotion is not enough, though. You also have to understand how emotions are used, both by yourself and others. That's the second skill in your armoury. Not everyone smiles when they are happy, or scowls when they are angry. Someone who is highly emotionally competent has a broad vocabulary of emotional concepts that are very flexible. They know how to impose meaning on smiles and scowls, frowns or vocal cues. They can take emotional signals from outside the world and from their own bodies and make sense of them.

This ability isn't innate – we're not born knowing the difference between feeling or seeing someone overwhelmed, worried, elated or ecstatic. It's a language that has to be taught. In an attempt to do just that, a programme called RULER was developed in the United States. The acronym stands for Recognising, Understanding, Labelling, Expressing and Regulating emotion. It teaches children and adults to interpret physiological changes in their bodies linked to emotions, name them, and learn strategies to use them appropriately.

It appears to have a tremendous impact on children's competence, and improves the relationship between teachers and students. One study showed a 10 per cent increase in academic performance, and another a 12 per cent improvement in classroom climate after just one year of implementing RULER. It is now used in more than 2,000 US schools as well as a host of other locations, including the UK, Australia, Italy, Spain, Mexico and China. Assessment of its impact on children and adults is ongoing, but if you're interested in using it in your family your best place to start is the Yale Center for Emotional Intelligence, where the programme was developed.

The third and final skill you need to hone is the ability to regulate your own feelings, once you've recognised and made sense of them. This ensures you properly analyse and appraise a situation, and that you conform to certain social standards. Again, this isn't something we are born with. As we develop, some of us learn ineffective strategies for doing it, such as avoiding emotionally charged situations or trying to shut down our emotions completely.

Anecdotally, this third pillar really resonated with me. I am definitely someone who avoids conflict. In the past, I have bent over backwards to not get into any kind of difficult conversation at work. I thought that was a good thing, preferring to go out of my way to get stuff done without causing a fuss. But research shows that this kind of attitude isn't the right one. People who address emotional situations directly rather than avoiding them have higher levels of well-being and are better able to cope with stress.

On discovering this, I have made more effort to be upfront with people, and in the short time since making this change I have found that it immediately goes a long way in decreasing anxiety, rather than increasing it. It might sound obvious that addressing a situation head-on works to improve your mental health, but it acts as a reminder as to how sometimes we can be blind to our behaviours until we give them some serious thought.

Regulating your feelings isn't just about facing difficult situations head on. There are several ways to improve this skill. One approach psychologists favour is 'reappraisal' – trying to put yourself in someone else's shoes so as to be more objective, and change your emotional response accordingly. When a team led by Ute Hülsheger at Maastricht University in the Netherlands taught this strategy to hairdressers, waiters and taxi drivers, they found that it resulted in more tips. They hypothesised that re-appraisal helped them to display more authentic positive emotions, and that is rewarded by customers.

Yet again, mindfulness is another promising approach – we've heard rather a lot about this in various parts of this book already. In a separate study, Hülsheger randomly picked members of a group of sixty-four employees to receive mindfulness training, and monitored them all over ten days. Those who got the training reported more job satisfaction and less emotional exhaustion. The idea is that when you just see emotions as they are, as thoughts and sensations, you gain a sense of perspective and the 'hot' aspect of the emotion dissolves.

Just like mastering any language, it takes time and practice to get to grips with our emotions, but making the effort is worth it, because the proponents of emotional intelligence were right about one thing – being emotionally fluent really does bring great benefits.

TOP TIPS FOR WISING UP

🔧 Think carefully about the framing of statements to work out whether how something is phrased is possibly misleading you about the content.

🔧 If you want to know how good you really are at something, err on the humble side when judging skills outside your main specialism, take a professional test, or ask a close friend for feedback – they're often more accurate than your own self-knowledge.

🔧 Get emotionally wise by practising 'reappraisal', which is where you put yourself in someone else's shoes so as to be more objective, and then changing your emotional response accordingly.

🔧 Don't shy away from difficult conversations. People who address emotional situations directly have higher levels of well-being and are better able to cope with stress.

Learn to identify the unconscious biases that can lead you to a bad decision, and get to understand them simply by asking yourself 'Why do I think that?' Regularly consider the opposite of what you had just been thinking, actively challenge your assumptions and intuitions and look for alternative hypotheses.

11

HOW TO GET AHEAD IN LIFE

TO WANT TO be successful in what you do – and where possible to enjoy the fruits of that success – is a natural human striving. Self-help manuals thrive on the desire of many people to get ahead in the world of work, providing all kinds of advice on the behaviours we need to adopt to maximise our potential and move onwards and upwards.

The genre is often accompanied by a cult of celebrity. We love successful people, and often turn to the likes of Elon Musk, Richard Branson or Sheryl Sandberg to identify the secrets of success that we can apply to ourselves. Surely there must be some kind of rule, personality trait or way of working, we think, that we can emulate to achieve greatness? I hope I won't be disappointing you at this stage if I say this isn't that kind of book. I'll get on to why, in my opinion, any self-help manual based on individual anecdotes about successful people, and the character traits that led them to that success, is bound to be flawed, in a moment.

My argument is based on science – statistics, in this case. You'll have noticed that science is my bag. I don't want to overstate what science can do for you – we live in a messy human world where set rules don't always apply. In what follows in this chapter, we discuss what science says about how you can cultivate traits to help lead you to greater success as you interact with people and attempt to forge a path for yourself in the world. It is itself messy, and it's incomplete – but whether your goal is to nail that interview, work more productively, think more creatively, or just darn well not get lost as you progress towards your goals, I think it's a whole lot more instructive than any celebrity CV.

HOW TO UNDERSTAND
THE SECRETS OF SUCCESS

Whenever I hear about celebrities or icons of business, or read articles about why they, above anyone else, have reached the top of their game, I think of Derren Brown, the English illusionist. In 2008, he offered the nation his foolproof system for predicting the winner of any horse race, in a TV show called *The System*.

To prove his point, Derren texts single mum Khadisha the name of the horse that will win a specific race. He's right, she wins. A few days later Khadisha receives another text with another horse. She bets even more money and wins again. Again and again she wins, betting more money each time. On her fifth race, a TV crew accompanies her to a real track and yet again her horse comes in to win from behind, having overtaken the two leading horses who both throw their jockeys off at the final fence. By this point she is convinced, as we all are, that Derren must have some secret method for predicting the winner. Khadisha meets Derren in person and places a £4,000 bet – her entire savings – on the sixth race.

Then Derren reveals his incredible secret. She's not the only person who has been receiving his texts. There are 7,776 people who were given one of six horses to back in that first race, leaving 1,296 winners. Each of those was given one of six horses to back in the second race. This went on and on until he was left with six individuals, including Khadisha, who had won four times in a row. Each was assigned a camera crew and a horse which they backed in the final six-horse race. The other five lost. Khadisha had won – every time – through chance alone.

A fact proven by Khadisha losing all her money on the sixth race – a race she had a one in six chance of winning. (In a twist, Derren later reveals that he placed his own money on the horse that actually won and gives her his £13,000 winnings.)

Derren's trick eloquently represents something scientists call 'survivorship bias', or what is sometimes known as 'the tale of forgotten failures'. When we focus on successful people and try to identify the characteristic that separates them from everyone else – Steve Jobs's temper or Sandberg's high IQ – we come to the conclusion that it must be these things that helped them get where they are. We forget that there are thousands of others with a strong temper or a high IQ who weren't successful.

Survivorship bias is the Achilles heel of any self-help manual (by the way, it didn't end well for Achilles, either): individual stories, or the bit of the story that presents itself to you or seems worth telling, is not the whole story. Another famous anecdote that eloquently demonstrates this is that of Abraham Wald, a statistician who was asked to work out how to protect planes better during the Second World War. His initial approach was to study those planes that returned from combat, thinking he would reinforce the areas where they had been hit the worst.

Then he realised that the planes that could provide the best evidence were those that hadn't made it back. Planes that were hit and arrived back clearly didn't need reinforcing in those areas – they could manage without. It was more likely that the parts that hadn't been hit were the reason they survived, and those were in fact the areas that needed reinforcing.

Before I am accused of bias, I should add that similar biases crop up in science, and they can be a huge problem. Studies that show a negative effect or a failed experiment have much less chance of being published in the biggest journals. Science is a human endeavour too, and like our species as a whole science journals have become accustomed to focusing on the few positive results and ignoring the many negative ones that might be just as helpful in directing us towards the next scientific advance.

Considering what you can't see is difficult. I'm not saying that a mercurial temper and a high IQ can't get you somewhere in your career. But we'll never know until we study that behaviour or trait within a much bigger population of people, some of whom made it and some didn't. Using informed studies rather than individual idols is the way forward when trying to plan your own path to success.

It turns out that a lot of hard-working traits that are traditionally prized in the office – staying late, missing lunch or sucking up to the boss – are working against you. Sometimes finding success in the workplace might be as easy as changing your hours, your handshake and even the way you dream. And despite your years of experience, your chances of that next promotion might be dashed by something as simple as the timbre of your voice or the symmetry of your cheekbones.

Yes, it's arbitrary and unfair. But while being rich, lucky or beautiful may help you to overcome the odds, you don't need them. A bit of social psychology can go a long way, too – and that starts with getting your foot in the door.

HOW TO SAIL THROUGH
AN INTERVIEW

One thing you can be sure of when you walk into an interview is that you're not there just to be tested on what you know. The people sitting in front of you are probably already aware that when it comes to technical skills and qualifications, you tick the boxes. What they're dying to find out is what you're like as a person – whether you'll fit in, whether they can trust you, how you're likely to behave at the office party. From now on, it's all about chemistry – or, more accurately, psychology.

So while I can't prep you with the answers to their questions, I can help you acquire a few tricks to give you the best chance of success. You might be thinking, just be myself. But that's only a sure-fire winner if you're the chief executive's cousin. A better strategy is to exploit some of the psychological shortcuts interviewers unconsciously use when deciding whether or not they like someone. We all use them when meeting someone for the first time, and research shows that interviewers rely on these more than rational analysis when assessing a candidate. I'm not advocating wholesale deception, just a bit of fine-tuning to help pitch things in your favour.

Not that you can control everything. As humans we are all prone to unconscious or implicit biases: if you haven't passed that way already, you can read about some common biases that plague human thinking in Chapter 10. Our implicit biases can be based on race, sex or a whole host of other characteristics – in any situation unreasonable and unfair, but in an interview situation doubly so. An increasing number of companies are offering training programmes to encourage diversity in hiring

265

workers and to make interviewers more aware of these unconscious biases, realising it's in their own interests, too.

Those lessons will hopefully be learned. Meanwhile, there are other unfair biases that we can at least begin to steer against. One universal one in performance at interviews concerns attractiveness. One reason for this is what's known as the halo effect: people assume that someone who scores highly in one character trait also scores highly in others. Social psychologist Richard Nisbett demonstrated that the thought process behind the halo effect is almost entirely subconscious. Our bias towards beauty probably has a deep evolutionary origin.

Ze'ev Shtudiner of Ariel University in Israel and Bradley Ruffle of Wilfrid Laurier University in Ontario, Canada, sent out 5,312 fictional CVs in pairs to 2,656 real job openings, across ten fields including finance, engineering, computer programming and sales. One of each pair of identical CVs had a photograph of either an attractive or plain male or female applicant, while the other had no photograph.

The team's fictional handsome men received 50 per cent more callbacks than men who didn't include a picture, and twice as many as plain-looking male applicants. Pretty men certainly benefited from their looks, but not so for women. Applications from women who didn't include a photograph were most likely to be picked up by recruiters, and plain-looking women were favoured over attractive women.[1] The team reckon this might have something to do with the fact that the majority of people in HR making the cut are women themselves, and may have felt some kind of threat from the better-looking female applicants. It all suggests that unless your job requires it, or you happen to be a particularly attractive man, it might be best to avoid attaching a photograph to your CV.

In an ideal world we'd all interview from behind a screen. That's unlikely to happen any time soon, so instead use the emphasis on attractiveness to your advantage as far as you can and make the most of your appearance: most interviewers are mugs just like everyone else when it comes to the subtleties of social psychology.

Wearing a made-to-measure suit can help if you can afford it – men wearing such suits were judged as significantly more confident, successful, high-earning and flexible than those wearing an off-the-peg suit of the same fabric. Those who can't afford tailored suits should at least ensure that theirs fits properly. Women might want to consider the length of their skirt and how many buttons they do up on their top. In one study, women who wore a skirt above the knee and unbuttoned an extra button on their blouses – even when the rest of their outfit was very conservative – were regarded as less employable and less confident.[2] Other studies suggest tattoos and body piercings are generally frowned upon, unless of course they're particularly relevant to your industry.

But interview success is by no means all about physical appeal. Be nice. It sounds obvious, but research shows that people take just one-tenth of a second to form judgements about a person's attractiveness, likability, trustworthiness, competence and agreeableness. Having more time to deliberate doesn't change our opinions, it only increases our confidence in them.

Unfortunately, it's not clear how accurate these snap judgements are, so do your best to walk into the room looking upbeat and friendly. Keep it up for at least half a minute. One study found that untrained observers who watch a video of the first twenty to thirty seconds of a job interview were astonishingly accurate at predicting whether the applicant would be offered the job. That doesn't mean the observers were especially good at picking good candidates. It means the interviewers, despite being fully trained, still go with their initial gut instinct.

Unless you're still in the midst of social distancing, shaking hands with your interviewer is probably the second opportunity you'll get to make an impression. Seize it – but not too hard. Several studies have found that people unconsciously equate a firm handshake with an extroverted, sociable personality – and that's more likely than a shy disposition to please an interviewer. What's more, a handshake can set the tone for the entire interview because it's one of the first non-verbal clues an applicant gives about their personality. People with firm handshakes have been shown in numerous studies to be more likely to be hired than those with a limp handshake.

You might think about making your hand warm, too. People who have been holding a warm cup of coffee were kinder towards others and viewed them in a more flattering light than those who had been holding a cold cup. This may not be as absurd as it sounds, since temperature and empathic feelings form close connections in the brain. Although this hasn't been tested in a real-life interview scenario, it suggests that an interviewer might look more favourably on an interviewee they've shaken hands with if their hand is warm.

Social niceties over, we're now getting into the meat of the interview. When we discussed small talk in Chapter 4 on making friends, we saw how mimicking is something we all do to empathise with those around us. Imitating the facial expressions, body language and mannerisms of the people we're talking to can also make them like you more. For instance, a study found that students rated a new soft drink more highly if the sales rep promoting it mimicked the students' physical and verbal behaviours.

However, this does come with risks in a job interview. Too much intentional mimicry doesn't work because of the divided attention it requires. It disrupts the flow and synchrony of an interaction, which then outweighs any positive feedback you get from the mimicry. And if you're caught doing it, it'll definitely leave a lasting impression, and probably not a good one.

No matter how successfully you've followed these rules, the interviewer will inevitably ask you all the things they think will help them decide whether to give you the job. But you know better: they're not going to decide so much on what you have to say as on how you behave. Remember: we tend to like those with whom we have something in common, as well as those who seem to like us.

So agree with the interviewer where you can, nod your head and smile. It may be difficult, especially if you've already formed a not so good opinion of your interviewer, or you're incredibly nervous, but ingratiating yourself really does increase your chances of getting the job – probably because if your interviewer gets on with you they will assume you share their beliefs and attitudes. And the halo effect applies here too: if an employer thinks you're likeable, they may also think you are intelligent, hard-working and competent. Which, of course, you are.

One more tip, if a potential employer asks when you can come for an interview, ask for a slot at the start of the working day, or just after lunch. This is when people tend to be in better moods and make more favourable judgements. This has been known to influence a very specific type of interview, and one of the few on which reliable data is available. When granting parole requests, one study showed that judges acquiesce to 65 per cent of the requests they hear just after their morning and lunchtime breaks.[3] As the day wears on – presumably as they get increasingly hungry and tired – the number of requests they grant steadily dwindles. Let's just hope that if you get that job, it doesn't feel like a life sentence.

HOW TO PERFECT
THE ART OF PERSUASION

So you've got your foot in the door. Now, at some stage you're going to want things. Whatever your desire or stage in life, the subtle art of persuasion can be a handy skill to have: just imagine that your honeyed words could automatically sweet-talk your boss into giving you a raise, for instance, or persuade your co-workers to change the way everybody does things. Unfortunately, the art of persuasion is just that: subtle, and notoriously hard to master. Done well though, it's almost impossible to resist.

Before you start, you should think about how to break down someone's resistance to persuasion. In fact, this may be more important than working out how to best get your message across. We are naturally suspicious of attempts to persuade us, particularly if we think we're about to be duped. Merely reminding people that they are vulnerable to manipulation, for example, showing someone an Instagram advert with a celebrity endorsing a product they clearly know nothing about, can make that person generally more difficult to persuade.

Perhaps that's obvious, but there's an important point to understand here. Resistance means that very persuasive arguments can backfire. People who successfully resist persuasion often become more entrenched in their wrong-headed opinions, and the stronger, more authoritative they perceive your attempt at persuasion, the more certain of their opinions they become.

It seems that if people resist good arguments presented by an expert, they conclude their own arguments must be even stronger.[4] This can lead us to a catch-22: it's good to have a strong argument, but if the person you want to persuade resists it, your message might be the trigger that enhances the very attitudes you want to change.

To overcome this deadlock, Richard Petty at Ohio State University, an expert on the unconscious factors in persuasion and resistance to change, advises trying to present positions that are closer to your target's views at first, then move them towards your goal a little at a time. You could also try charming them by boosting their self-esteem – when people feel good about themselves, they are more open to challenging messages.

Once you've got that balance right you might want to add 'framing' – a favourite tactic of spin doctors – to your argument. Framing is about leading people to think about an issue in a way that is advantageous to you. Say I wanted to sell you a new type of vegetable 'broccolulu' to replace your normal purchase 'cauliflop'. I could tell you all about the benefits of broccolulu to your heart health, but I'd be better off mentioning the damaging effects of cauliflop on your cholesterol. A large body of research shows that negative information frequently has a more powerful influence than positive messages.

Next it's time to support your argument. When you're thinking of reasons that your argument is the right one, it intuitively feels like the more arguments you can call on, the more persuasive you'll be. Wrong. Several studies suggest that the more reasons people are asked to come up with in support of an idea the less value they ascribe to each. The result is that asking people to think of all the reasons why this is a good idea is likely to backfire, and may serve to harden their original views.

For instance, in one study, students were told there was a plan to introduce new exams into their course – an unwelcome prospect. They asked half the group to produce two reasons why this was a bad idea and the other half eight reasons. On average, students who supplied two arguments against the proposal were subsequently more opposed to the exam policy than those who gave eight. It's generally easier to remember two reasons for believing in something than eight, and the ease with which we can summon our thoughts affects how much confidence we place in them. Getting people to generate just a few positive thoughts about your idea might be your best bet in convincing them it's the right way to go.

Finally, when preparing your argument, think about how it's delivered. Successful persuasion seems to have as much to do with how you say something as what you're saying. And the less time you're allowed to think about what someone's saying, the more the style of delivery matters.

When given a transcript of a sales pitch that described a scanner, people whose transcripts included hesitations like 'I mean' and 'ummm' were less easily convinced that this was the scanner that was worth buying – even when it was a better scanner at a lower price. People who didn't get given enough time to read it through properly were even less convinced by the sales pitch that contained hesitations. If you can't pay attention to what someone is saying, you pay attention to how they say it.

So if you want to be persuasive, frame your argument correctly, condense it down to a few salient points, don't try pushing things too far, too quickly, and try not to stumble or pause. And for goodness sake, don't give your listeners time to think about what you're really saying. Which brings us nicely on to our next subject.

HOW TO AVOID
GETTING INTERRUPTED

It has happened to all of us. You're in the middle of an important point at a crucial meeting, or just telling an amusing anecdote, and someone butts right in. You may jump back in to finish what you are saying, indignantly stammer a few more words or quietly fume while the interrupter takes the floor, but the moment has passed: your eloquent point is lost, your story garbled. Such scenarios are a regular and particularly annoying aspect of our working and social lives. Perhaps you're conscious of interrupting others from time to time (well done you for being so self-aware!). Maybe you don't think you interrupt, but know someone who does.

Men often get the blame, and stereotypes would also have us believe that people from some countries are more likely to jump in than those from others. There may be truth to some of this, as we'll find out, but a closer look at interruptions reveals why you might want to be a little more forgiving of the big mouth who stole your moment. But I'll give a few tips on how to avoid it happening so much, too.

Firstly, are men really to blame? The idea for this originally seems to have come from a study in the 1970s. It showed that, in covertly recorded conversations between men and women in the United States, men cut in a whopping forty-six out of forty-eight times.[5] And a 2014 study found that men and women both interrupted women more than men.

Before we stick it to the patriarchy, though, a word of caution. Interpreting these results is difficult. These studies tended to count all overlapping speech, but sometimes people overlap and it's not particularly interruptive at all. It was also acknowledged by these researchers that it's hard to know whether men were interrupting more because of their gender or their status, with men in the study more often holding positions of power.

To tease these subtleties apart, researchers at Northwestern University Pritzker School of Law in Chicago headed to the US Supreme Court. Here, nine justices must reach a decision, and the ability to dominate the floor can determine the fate of a case. By documenting hearings over several years, the researchers clearly established that women are interrupted at a markedly higher rate than men, regardless of seniority. Male justices interrupted female justices three times as often as the reverse. Female justices were also interrupted three times more than their male colleagues by male lawyers arguing their cases, even though clear rules forbid this.[6]

Interruptions go beyond gender. Italians are famed for talking over one another, Japanese speakers reportedly leave long gaps between each person's turn in a conversation, and the joke goes that if you offer Swedish visitors a cup of tea, you might wait a whole minute for a response. The English are supposed to have finely honed their skills of non-interrupting, relying on grammar and intonation rather than a distinct pause in conversation to know when it is their go.

Unsurprisingly, these somewhat offensive stereotypes turned out to be false. When Australian researchers analysed hours of natural conversations between people speaking ten languages over five continents, the English speakers took about 240 milliseconds between speaking turns, Danes waited nearly half a second and the Japanese were quickest to respond, jumping in after 7 milliseconds.[7] The ability to take turns without talking over each other or waiting too long seems to be universal across languages, geography and culture. The stereotypes instead might come from the fact that we are so finely attuned to our own timings in conversation, we feel tiny differences to be much longer, or shorter, than they actually are. And this leads to interruptions.

Culture also makes a difference to the content of the interruption. Chinese, Thai and Japanese speakers tend to interrupt each other more often with cooperative speech, such as agreeing with what was said or adding assistance to an idea, rather than intrusions which attempt to steal the floor or change the subject.[8]

Men and women also use different types of interruptions when talking to their same-sex friends. Women interrupt each other more than men do, but their interruptions tend to be to agree and build on the point being made, rather than argue or change the topic.

So how do we attain conversational harmony? Firstly, we need to take all of these findings with a pinch of salt. The science of linguistics is incredibly complex, with many subtle differences in the way we talk to one another, and few large studies dive into it. The solution, perhaps, is to become more aware of how conversations work so we can overcome our instincts to quickly jump to conclusions about the intentions of others. People who respond quickly aren't necessarily pushy, people who respond slowly aren't docile, and someone may leap in to show interest, not to take over. And when someone speaks out of turn, remember they might just be experiencing a language-induced time warp.

Of course, sometimes when you need the floor, you might just have to raise your voice and jump in – it's part and parcel of the way our communication works. Some research suggests that this strategy is viewed more negatively when women do it than men, but it may be a trade-off women have to make. Once you have the floor, speaking quicker than feels natural can help prevent further interruptions. An old study from 1983 suggests that when women physically leaned into a conversation they were less likely to be interrupted, and more likely to be interrupted if they weren't looking at the interrupter directly in the eyes.[9]

Finally, if there are men – or women – who are interrupting consistently and uncooperatively, try pointing it out to them. Since the Supreme Court study was published, Chief Justice John Roberts is thought to have given the floor back to women more often than before – although this hasn't been officially quantified.

HOW TO STOP PROCRASTINATING

Douglas Adams did everything humanly possible to avoid the daily drudgery of plonking down at his desk and pounding out his novel *The Salmon of Doubt*. The eccentric British writer soaked for hours in the bathtub, frittered away entire days in bed and dreamed up ever more fanciful excuses for his exasperated editor. When he died in 2001, he had spent a decade on the book without even a complete first draft to show for it. Adams, whose works include *The Hitchhiker's Guide to the Galaxy*, was a poster boy for procrastinators everywhere. 'I love deadlines,' he once quipped. 'I like the whooshing sound they make as they fly by.'

You might think you do your best work under pressure. I've certainly claimed that on numerous occasions, but studies suggest we're deluding ourselves. Regularly delaying tasks you know you should work on immediately doesn't just prevent you from achieving your full potential: it can be expensive, bad for your health and could even endanger your life and others.

When researchers took a look at the self-employed, for instance, they found that individuals who leave the preparation of tax returns to the last moment make errors costing them £400 per return on average – certainly no pay-off there. And when students were given various assignments that had to be submitted online by a specific time, those who scored low on a questionnaire designed to measure their tendency to procrastinate and who worked at a steady pace had an average grade of 3.6 out of 4. Not so those who scored high on the questionnaire, whose grade average was just 2.9.[10]

And it's the same wherever you look. International studies of procrastinators from across Europe, the United States, Canada and Australia show that those who tend to postpone things are also less likely to get annual medical and dental check-ups, and exercise regularly. They suffer from more stress and illness and digestive problems. And the more serious the procrastinator, the less likely they are to do things to prevent home accidents, from owning a fire extinguisher to making sure electrical appliances are safe.[11]

Someone who has researched this in great detail is psychologist Piers Steel at the University of Calgary in Alberta, Canada. He has spent countless hours poring over the results of hundreds of studies on procrastinators, to come up with four factors that are most strongly linked with procrastination. They are: how confident a person is of completing a particular task successfully; how easily distracted an individual is; how boring or unpleasant the task is; and how immediate the reward for completion will be. The more uncertain of success or easily sidetracked you are, the more likely it is that you will put off an assignment or chore. Conversely, the more pleasant the task and the more immediate its payback, the greater the chance you will get on with it quickly. He also discovered that men postpone things slightly more than women, and the young tend to loiter over tasks considerably more than seniors do.

So what can you do to stop yourself dragging your feet? Making a task appear less unpleasant or more immediately rewarding is an ideal start. A treat after you've completed 1,000 words, for instance, or booking yourself a holiday for the end of a big project, come to mind. Next, you want to minimise the distractions around you – switch off your phone and computer alerts. Get a good night's sleep: as we know, tiredness is the enemy of motivation. Make commitments that you cannot get out of, or at least will come at a high cost of embarrassment or potential job security. And remind yourself that all of this really should lead to a better outcome – whether that's a higher grade or more exacting prose.

Finally, you could copy Douglas Adams, who outsourced his motivation to others. His editor once booked him into a hotel room and stood guard over him until he finished a promised manuscript. There's nothing like a watchful eye to hold your feet to the fire.

HOW TO BE MORE PRODUCTIVE

Douglas Adams excepted, if you want to get on in the modern world, it often seems like you need to go full-on, optimising productivity by skipping lunch and fielding emails 24/7. Busyness has become a powerful and aspirational status symbol. When researchers asked volunteers in the United States to assess the status of fictional individuals from short descriptions, anything indicating the person was extremely busy resulted in higher estimations of the person's importance and achievements.[12]

But are they right? Does being busy really lead to success? Intriguingly, when the experiment was repeated in Italy, the effect was reversed – people described as busy resulted in lower estimations of their achievements. So who is right?

You suspect that Adams would have been a friend of the Italian concept of *dolce far niente*. It essentially means the 'sweetness of doing nothing' (the English version of the phrase tends to be a little more robustly expressed). And surprise, surprise, it seems like it is the winner's formula. In 2018, researchers examined data from 52,000 employees across Europe. They found that people who worked intensely for long periods of time, working at high speed or to tight deadlines, scored lower on measures of mental and physical well-being.

OK, you might say, but did they benefit from this sacrifice, career-wise? No. These staff were also less likely to be promoted, or feel satisfied and secure in their jobs. That might sound counter-intuitive, given everything we hear about hard work and success, but escape from work is essential to maintain the cognitive capacities required for productivity. Skipping breaks is the best way to deplete energy reserves, increase stress and depress your productivity. When the Draugiem Group, an IT company based in Latvia, tracked the behaviour of its employees, the top 10 per cent most productive people worked the same hours as everyone else, but took more breaks – on average a seventeen-minute break after fifty-two minutes of work. At the very least, you should try to stand up and move around for thirty to sixty seconds every twenty to thirty minutes, and focus on something more than six metres away to rest your eyes. And definitely try to take an extended break of at least ten minutes every hour or so, because we have only a limited capacity to concentrate for longer.

Taking a break isn't just good for your cognitive muscle, but also for your overall health. Extended periods of inactivity, as we discovered in Chapter 6, increase our risk for all sorts of diseases. Adults who sit for one or two hours at a stretch over a regular period of time have a significantly higher risk of dying prematurely than those who spend the same amount of time sitting, but who get up and move every half an hour or so.[13]

For maximum effect, there is a better and worse time to take a break during work. Focus your free time on the afternoon. An analysis of more than 500 million tweets across eighty-four countries showed that our mood follows a predictable path over the day: positive first thing, and then gradually souring.[14] Positive mood aids creativity, decision-making and working memory, suggesting that tasks requiring those skills − running meetings, devising strategies, plotting hostile takeovers or starting a new craft project − are best tackled early in the day.

Our cognitive performance follows a similarly predictable path, with most people at their best in the first half of the day, peaking around midday.[15] There follows a dip, and then a second peak, but then you're on the way down. The afternoon slump is not an urban myth, but a natural feature of our body clock. We encountered adenosine, a chemical tracker of how long we've been awake, in Chapter 7 on how to get a good night's sleep. It starts to exert 'sleep pressure' as it builds up in your brain. So all those mundane tasks you probably do first thing, like checking your emails and filing your expenses, may be best saved for your dog-day afternoons.

Of course, a complicating factor is that we all have what's known as a natural chronotype, which might not be in synch with our working hours. Around 20 per cent of people are 'owls' meaning they naturally sleep late, wake late and function better later – that's definitely my natural chronotype. Another 20 per cent are 'larks', at their chirpy best first thing in the morning. The rest of us lie somewhere in between. When we perform work in hours that we'd naturally be sleeping it can be very unproductive – studies suggest only 45 per cent of people have zero conflict, while 10 per cent have a conflict between the times of day they have to be active and the times of day they would want to be active of more than two hours.[16] These are probably owls, forced to work when their biological clock is screaming for them to stay asleep.

Owl or lark, if you find yourself in a slump, try to take a power nap – just twenty minutes can increase cognitive perform-ance and alertness (you can find out exactly what kind of nap is best for you in Chapter 7, too). Or if that's not possible, get some exposure to natural light or have a cup of coffee: this will help to block adenosine, reducing sleep pressure and mimicking the effect of a nap.

It's not easy, especially when you've got looming deadlines and a hard-nosed boss breathing down your neck, but by being as strict with your breaks as you are with your working hours, you will find you have more fulfilling and productive days. But above all don't forget to get a good night's sleep – as we dis-covered earlier, the perfect day begins with a perfect night.

HOW (NOT) TO CONCENTRATE

Wake up! You're sitting in an important meeting but it's been an hour and you just can't seem to take in what's being said. You have a deadline for your tax self-assessment, but no matter how hard you try to focus, you just can't. Your exams start in less than a week and the words are blurring on the page. The clock is ticking, but the sun is shining and, oh, is that a barbecue you can smell?

Apologies if you didn't get that perfect night's sleep last night, but losing concentration is something that we all encounter daily – and it can be a serious problem. On a normal day, studies suggest that we spend as much as 50 per cent of our time thinking about anything except what we should be doing. Any number of studies link failing to concentrate with unhappiness, stress and a lack of success.

Recently, however, psychologists have been having a rethink. If we spend so much time in a state of reverie, it's probably not a mistake. And in fact, while you might consider your mind-wandering a bad habit, it turns out there are several kinds of mind-wandering and some of them might be a key weapon in your cognitive arsenal, if you know how to use them.

To master mind-wandering you first need to understand a little about what's happening in your brain when you try to concentrate. Broadly speaking, we have one attention system that alerts us to anything that suddenly stimulates our senses – a loud noise, a tap on the shoulder, a WhatsApp notification. These sudden alerts might be distracting, but it's a skill that evolved for a reason – it's no use focusing well enough to carve the perfect spearhead if you get eaten by a lion before you can use it.

So ingrained is this alert system that ignoring these disturbances is really difficult – and as we saw in Chapter 8 on habits, dealing with them can border on the obsessive. The solution is obvious. Shut them out: switch off your phone, disconnect your WiFi, block unpredictable noise.

But even these measures can't stop a second brain system from leading you astray. When we concentrate on a task, we use our 'executive control network', a set of brain areas that are responsible for goal-oriented thinking and controlling impulses. But this is in a constant tug of war with the 'default mode network', which fires up when we think about nothing in particular. The default mode network does our mental house-keeping – flicking through memories, forward planning, filing away bits of information, but it's also the brain region that is most active when we daydream. To stay on track, we need to keep the volume of this chatter to a minimum.

The problem is, the brain seems to find mind-wandering much easier than concentrating. Neuroscientists have long debated why. One answer might be because the default mode network is highly connected to lots of parts of the brain, which allows us to flit between different thoughts with little energy. The executive control network, on the other hand, is more sparsely connected, so perhaps needs more effort to shout over the noise.

All this would be rather hard to account for evolutionarily if mind-wandering were a fast track to failure. The reason we thought it was for a long time is because researchers investigating the subject assumed that volunteers asked to do a boring task in the lab would try their hardest to concentrate until mind-wandering unintentionally took over. What they failed to consider is that sometimes we intentionally let our mind drift to more appealing topics, especially when our main task is dull or of no use to us.

When scientists started interrupting people during tasks to ask whether their minds were wandering and whether it had happened intentionally or not, they found that more than a third of the time the mind-wandering was intentional.[17] In some studies, people seemed to daydream purposefully more than half the time.[18]

This brings us to a rather different thesis about what's going on. A lot of the time your executive control network isn't losing its grip on the default mode network when your mind wanders: it's actually in charge of the whole experience. The distinction is important: in the past, mind-wandering has been linked with symptoms of ADHD and OCD, both conditions in which a lack of control over certain behaviours can interfere with getting things done. But newer studies show this is true only of unintentional mind-wandering, not the more directed kind.[19]

Context is also important. When our mind wanders during highly demanding tasks it is more likely to cause memory lapses, bad performance and risk-taking. But if your mind wanders during less arduous tasks, studies show that it is associated with a better memory of the events you're working on, improved creativity (an important word we're going to hear more about), as well as more patient and prospective decision-making, as opposed to impulsive choices.[20] This all seems to be thanks to allowing the brain to forge connections between pieces of information we don't link up when we are too focused on one thing. Which means there is a good time to let your mind off the task at hand, and a worse time.

Now let's think about the content of your daydreams, because that's significant too. Thoughts about the past are much more likely to lead to low mood and motivation than those about the future. In fact, future-related mind-wanderings actually seem to boost our mood and motivation, even if those thoughts are about flunking out of college or not getting that promotion.[21]

Taking charge of your thoughts might seem difficult when you're struggling to focus at all. But there are some tricks you can use to veer your mind towards the most useful kinds of wandering. If you've got a bit of time to dedicate to this particular habit, you might want to think again about mindful meditation. Some small studies have shown that people who are more mindful have a greater ability to deliberately mind-wander. Paul Seli, a psychologist at Harvard University, believes that we could all learn a thing or two from them, by learning mindful meditation – something we have encountered already – to turn our unintentional mind-wandering into the more helpful, deliberate kind.

Meanwhile, there are simpler strategies available. Rather than wait until you really need to concentrate, try practising intentional mind-wandering when relaxing. Studies suggest that this helps people keep on track when they really need to concentrate.[22] Another tip is to bribe yourself with a well-deserved reward. Make this a good one, which you get at the end of a task, rather than smaller treats throughout. These smaller rewards didn't make any difference to people losing focus during a boring task, but the promise of a larger final reward kept them alert until the end.

Doodling might also come to your rescue. When people are forced to listen to a tedious voice, those who were allowed to doodle remembered more afterwards. But the devil is in the detail: doodling stuff related to what you're trying to remember was helpful. It qualifies as intentional mind-wandering, which helps you to focus on the task at hand. Make your artwork too elaborate though and the whole thing seems to backfire.

Finally, as I've said before, get some sleep. A lack of zeds reduces our ability to resist internal and external distractions. In fact, research suggests that if you have an hour spare before an exam, a nap could be a more effective use of that time than revising, given its importance in consolidating previously learned memories.[23] And who needs a better excuse to have a daytime nod? Not least because it helps you with what we're about to discuss . . .

HOW TO BE MORE CREATIVE

About fifteen years ago, I was sitting at my desk trying to work, when from nowhere came a fantastic idea. I told my colleagues about it. 'Why doesn't someone rescue all the food that is rejected by supermarkets for being too big or small or a weird shape and sell it to those who don't mind a bit of lop-sided food?' I said. I never did anything more about it – and a decade later, Wonky Vegetables, Oddbox and the Ugly Company are all very successful at doing just that. Oh well. I'm still going to take credit for that Aha! moment of inspiration. We all have them, and they seem to come from nowhere. They come to me at the weirdest times, and never when I really want or need them. I would love to have more (and to actually act on the good ones).

Our ability to be creative, to think outside the box, or come up with new ideas, isn't often reflected in standard tests of intelligence. In fact, there's no consensus as to whether general intelligence and creativity are linked. There seems to be a low threshold of intelligence, as measured by IQ, that is needed for creative potential. Once this is met, however, personality factors – such as 'openness to new experiences' – are much more predictive for creativity.

That's actually good news for those of us who would like those moments of creative insight to come a little more easily, and a little more often – there's nothing stopping them doing so. And it turns out there are several ways you can help your unconscious do its work. In terms of lifestyle: getting enough sleep, and especially dream-containing REM sleep, seems solidly correlated with more creativity (see Chapter 7 for more on what sort of sleep and when). So perhaps there's something in the cliché of the lazy genius who sleeps till noon.

But don't be lazy all the time: if you're looking for a quick burst of creativity, a walk or similar exercise is a good way to unleash it. One study, the idea for which came when the lead researcher was out walking, found that people came up with more uses for everyday objects when walking outside or on a treadmill than when seated. And people continued to be more creative afterwards, suggesting that a saunter before a brain-storming session is a good idea.

As far as the process of creativity is concerned, in 2019 researchers at Columbia University, New York revealed that Aha! moments occur when enough relevant information has accu-mulated in the unconscious to trigger conscious awareness of a decision. The point at which this critical threshold is reached will vary depending on the task. However, some people seem better at achieving it than others. There are a couple of possible reasons for that. Studies suggest variously that creative insight is driven by one of two very different states of mind: concentrated focus and daydreaming. Intrigued by the contradiction here, Jonathan Schooler at the University of California, Santa Barbara decided to test them head to head. He found that focused thinking actually undermines inspiration unless you are using an overtly analytical approach to solve a problem. By contrast, as we hinted at in the previous section, letting your mind wander, after taking in information, cultivates creative insight across most other tasks.

If you want more Aha! moments, you must first scour some relevant material to give your unconscious something to work on. Then Jonathan Schooler recommends finding time for un-focused thinking. This is best done while you are engaging in an activity that's not too mentally taxing, such as walking, gardening or household chores. Try to disengage from spontaneous thoughts that are mundane, like thoughts about current concerns or plans for upcoming tasks, or thoughts merely replaying familiar scenes. People who experience more creative insight tend to report more bizarre imagery while mind-wandering, so try to emulate them. Engage with thoughts that are a bit more unusual or fantastical. Follow those thoughts through to the end, or extend them by asking playful, imaginative questions, such as 'What if x was different?' or 'What if x was reversed?'

Another way to tap unconscious inspiration is to modify your emotional state. There is some evidence that listening to positive background music, such as 'Spring' from Vivaldi's *Four Seasons*, helps people come up with more creative ideas. Researchers suggest this may be because it triggers the release of dopamine, which is associated with creative thinking. Christina Fong at Carnegie Mellon University, Pennsylvania, has found that simul-taneously experiencing two emotions that aren't typically felt together – such as frustration and excitement – encourages creative insights too. That might be, she says, because it signals that you are in an unusual environment, making you alert to the possibility of other unusual relationships. If so, then life will be more inspiring if you embrace change and novelty.

And though they say you won't find the answer at the bottom of your glass, it seems a little tipple may help you stumble upon it. A study exploring the influence of alcohol on creative problem-solving suggests a small amount of booze could help you find some answers. Forty men drank either a vodka and cranberry, adjusted according to body weight, or a non-alcoholic version. The subjects then took a test which involved linking groups of words with a single concept. The tipplers solved 38 per cent more problems than their sober counterparts and reached the correct answer faster. They were also more likely to say they hit on the answer with a 'sudden insight'.[24]

These insights were accompanied by a reduced working-memory capacity – a measure of your ability to focus on a specific task – which does support the idea that alcohol fuels your creativity by allowing your mind to wander, connecting disparate ideas. It is possible to get a little too unfocused of course – any more alcohol and the benefits disappear. Another word of warning is that this was a small study, and it's only been repeated on an equally small scale, so it's best taken with a pinch of salt (possibly without the tequila).

Finally, when talking about boosting creativity, there is 'flow'. This is a slippery concept, a sort of deep immersion characterised by automaticity – a sense, for example, that the novel you are working on is writing itself. Research suggests that flow comes when you 'turn off' conscious thought. Distractions will disrupt this process, and they are not conducive to daydreaming either. So, whether you are seeking flow, or trying to let your mind wander in the hope you will solve a problem, make sure you put your phone on silent and turn off your WiFi. Just this once, the answer to your problem isn't going to come from Google.

HOW TO AVOID GETTING LOST

It is autumn in New York City and a monarch butterfly is setting off on its 4,000-kilometre journey to a fir tree on a mountainside in central Mexico, where it will spend the winter, and where it subsequently arrives with pinpoint accuracy.

It is autumn in New York City and I turn out of a cafe, set off for my hotel just a few streets away, take a wrong turn and get utterly lost because the battery in my phone has run out. Perhaps it is unfair to compare my navigational skills to those of the legendary monarch butterfly, but however you look at it our sense of direction isn't up there with the greats. How many times have you come out of a shop on a busy street, turned left to head to the station and realised that the station is in fact to the right? Or perhaps you've wandered through some woods, and got a sudden fright when you can't find your way back to the main path? Or become immediately disorientated after your satnav loses signal?

I first got interested in navigation when I came across a woman with developmental topographical disorientation disorder. Or in other words, an utter inability to navigate. She had spent her life completely disorientated, and couldn't even find her way from her own kitchen to her bathroom. And while most of us will never get quite so lost, the tips and tricks that I learned from investigating this condition are now among those that I use the most in my life.

I think the easiest place to start is to think about how we get lost. To do that, we'll change scenes to the Canadian wilderness, perhaps a more extreme environment than most of us face most days – but bear with me, because the lessons apply to us all.

About thirty years ago, Ed Cornell, a psychologist at the University of Alberta in Edmonton, answered the phone. It was a police officer leading the search for a nine-year-old boy. He had gone missing from a campsite and his footprints suggested he'd headed towards a swamp a few kilometres away. The officer wanted to know one thing: how far do nine-year-old boys tend to travel? He asked Cornell because he and his colleagues had been studying wayfinding behaviour. But after pondering the question, they realised they knew very little about lost children. They weren't able to tell the policeman much. His response stuck with him: 'Don't worry, doc, we may get a psychic out here today.'

Since then, Cornell and others have discovered that although we behave irrationally when lost, we also share habits that might help people find us again. For a start, when we get lost, we tend to make things worse by moving around. In a review of more than 800 search-and-rescue cases from Nova Scotia, only two people had stayed put: an eighty-year-old woman out picking apples, and an eleven-year-old boy who had taken a survival course at school. The extreme stress of being lost, with its accompanying adrenaline rush and all-encompassing fear, makes it very hard to think rationally, notice landmarks or keep track of where we have travelled. When the mental performance of pilots and aircrew were tested while confined in an oppressive mock prisoner-of-war camp, their working memory and visuo-spatial processing – necessary for map-reading, spatial awareness and other navigation tasks – were so poor they resembled children under ten.

When under such stress people tend to do similar things. We are drawn to boundaries, such as the edge of a field, a forest margin, a line of pylons. In search-and-rescue operations, most lost people found alive end up in a building or along one of these margins. But our tendencies can vary according to age and gender. Children are less likely than adults to keep moving, people with dementia tend to head in a straight line and solo male hikers travel much further than any other kind of person.

When left to roam, children travel further than anyone

– especially their parents – thinks they do. Three or four times as far in some cases. When asked to head to a place they know well, they tend not to go directly to their target: they get distracted or take long, circuitous diversions.

So what does this tell us about how to avoid getting lost in the first place? Let's think about getting lost in the woods. The important thing is to stop moving when you feel like you're heading in the wrong direction, at least for a while. The powerful impulse to carry on walking in any direction to try to find your way is the cause of many tragic stories. If you don't have anyone around to help you, your next strategy should be to try to retrace your steps. This requires patience, which is difficult when you are scared, and psychologically challenging because it feels like you are moving further away from safety. Some experts recommend finding a landmark such as a large tree or recognisable outcrop and treating it as a hub of an imaginary wheel. Walk out and back along the spokes of the wheel while keeping your hub in sight until you find something familiar. Another tactic is to climb a hill or tree so you can spot more distant landmarks. Finally, if you're heading into the wilderness, always buddy up – two people will be less scared and more rational.

Now you know how to act in the wild, what about in the middle of Paris or on a country walk in gentler terrain? A lot of us get lost when visiting new cities or going for long walks on a weekend break; some people enjoy it, others not so much. If you're in the latter group, understanding a few things about how the brain forms our mental maps might help.

First, there seems to be no correlation between our wayfinding ability and scores on intelligence tests. Another reason our sense of direction has a chasm between individuals is due to the approach we take to navigating. Some of us tend to rely on route-based navigation, which entails remembering landmarks on a particular journey: we turn left at the bus stop, right at the yellow gate, and so on. This works well in familiar towns or regular journeys, but it would take far too much mental effort for new areas or when you're forced to go out of your way. Instead we also rely on 'mental mapping'.

To form a mental map takes a lot of communication within the brain – there's a specific region that identifies permanent landmarks, for instance. If this is lesioned or diseased you can take someone to the centre of Paris and ask them to identify a good landmark around which to navigate back to at a later time, and they're just as likely to choose a parked car as the Eiffel Tower. Then there are brain regions that help you identify boundaries, and cells that monitor how far you've walked and in what direction. There are areas of the brain that tell you about what direction your head is moving in, another that focuses on storing and retrieving memories of your location, and another that combines all this into a mental map that is constantly updated with any new information of note. The best navigators don't have the best memories, nor are they the best at understanding compass directions: they have the best communication between all of these areas. They also implement the best navigational strategy for each occasion.

While we're on the subject of differences between people, we should probably settle the persistent rumour that men are better at navigating than women. Of the hundreds of studies that test this theory, no one has come to an outright conclusion. Sometimes men and women perform equally well. However, in tests where men outperform women there are hints this might be down to a preference among men for using mental mapping, compared with a preference among women for using route-based navigation.

Some studies have also found that women are less likely than men to explore shortcuts. In ancestral times, and arguably still today, a woman who gets lost may be more vulnerable than a man, meaning any gains you get from the shortcut would be lower given the higher risk of unexpected threat. However, recently researchers have started looking at sex differences in navigation outside Western cultures, in small-scale societies, such as the Tsimane of Bolivia and the Twe of Namibia. Twe and Tsimane boys and girls are equally good at pointing accurately to distant locations, and at imagining being in one location and pointing towards another – two classic tests of navigation ability. Around adolescence, girls from both groups become increasing concerned about physical dangers, including the risk of getting lost.

However, although a difference in navigation emerges in Twe adults, it doesn't in the Tsimane. The Tsimane hunt and fish in dense, dangerous jungle and tend not to roam far, while the Twe live in open savannah. Unlike the Tsimane, Twe men have much larger ranges than women and travel long distances to visit partners; the society isn't monogamous. As a result they face greater navigational challenges, and gain more experience, which may explain the differences in adults. All this suggests that our navigation abilities have less to do with whether we are male or female per se, and more to do with environmental factors and personal experience. In the West, men tend to drive more than women, which might give them more experience at navigating, for example. Many studies also use virtual reality environments, giving an advantage to people who regularly play video games, which tends to be men.

Regardless of gender, there are ways to improve our everyday navigational abilities. First, pay attention to your environment, particularly any permanent and highly noticeable landmarks, and how they relate to turns you make. Make these connections as funny or as memorable as possible. For instance, a few years ago I visited Palma on the Spanish island of Majorca. While trying to find my way around the complex maze of back streets, I made sure to remember that my hotel was on the road with the corner store that looked like a sex shop due to a phallic red neon sign it had flashing in the window. I visited the city recently and that sign jumped out at me immediately. Even though years had passed, I could have relocated my old hotel with ease.

You should also regularly turn around and look behind you, a technique that some animals use to navigate more efficiently. Looking at what your environment looks like facing the opposite direction helps your brain piece together your mental map when you're travelling home. A final tip is that you can easily check which direction you're facing when the sun's out – by checking the orientation of your shadow, and using this to tell roughly which way you might need to head.

Of course, if the GPS on your phone is working, this is often what we use. But you shouldn't rely on it. Although the evidence is shaky on whether it is detrimental to our wayfinding abilities in the long run, it certainly impairs our ability to form a cognitive map of our environment in the short term, leaving us at risk of getting lost if the battery runs out or find ourselves with a broken phone.

But above all, the secret is practice. We know that people with a better sense of direction are more likely to explore an environment rather than rely on known routes, which helps build their cognitive maps and further improves their wayfinding. So if you want to improve your partner or your children's ability to navigate without GPS, let them practice. Just remember they'll probably go a lot further than you think.

TOP TIPS FOR GETTING AHEAD IN LIFE

🔧 Be nice and positive at the start of any job interview: people take just one-tenth of a second to form judgements about a person's attractiveness, likability, trustworthiness, competence and agreeableness.

🔧 Take a break of ten minutes every hour or so to maximise your concentration and health at work. People who work intensely for long periods of time score lower on measures of mental and physical well-being and are also less likely to be promoted, or feel satisfied and secure in their jobs.

🔧 To be more persuasive, try framing your argument. Negative information frequently has a more powerful influence than positive messages.

🔧 If you're finding it difficult to concentrate, try doodling stuff related to what you're listening to – it'll help you remember it at a later date.

🔧 When you're in a new city, make a mental note of landmarks and their relationship to one another. Turn around and look behind you every so often. This helps your brain create the mental map you need to stop yourself getting lost.

12

HOW TO AVOID
BEING TOO PERFECT

AS I SURVEY MY home-office-cum-playroom around me, with its jumble of wires and computer equipment, piles of paper and marker pens, and the mess of toys, sometimes it's hard to avoid my heart sinking a little. I've talked a lot about the circumstances of writing this book during the coronavirus lockdown, and I mentioned at the beginning of Chapter 1 the stresses I felt from the competing priorities of work, family, pregnancy – and not least writing this book.

I hope I've at least partially succeeded in one of my goals of the past few months. After everything we've come across in this book, you'll hopefully be on a path to being healthier, happier, smarter and more relaxed.

'Relaxed', huh? That's always a difficult one. Because there's a danger with all this self-help stuff: that in our striving to become better people, we actually make ourselves more stressed and worse off than if we'd just gone with the flow. It's something I've felt I've wrestled with in the past few months, and where my knowledge of the science has been a real help to me. So in this final chapter I'm here to throw a little cold water over your endeavours, and perhaps even some of my advice, to just rein you in a little. It's as well not to take things too far. Sometimes the best strategy is to let go, accept imperfections and simply make the best of things that we can. In all things, self-improvement not excluded, moderation is key.

HOW NOT TO BE PERFECT

We've all heard it, or perhaps said it ourselves: 'I'm a perfectionist.' Sometimes it's to give an answer to the laziest interview question in the world: 'What's your worst habit?' At other times it's to excuse pedantic behaviour or a genuine desire to make something, well, perfect. This hankering for the highest standards might have been why you bought this book in the first place, and hopefully a lot of what you have read will be the driver of a more successful, healthier, happier you. But for some people, diligence and motivation are habits that can shift into perfectionism, a trait that can have dangerous consequences.

Perfectionism has increased significantly over the past three decades, at least in people in the Western world, a recent analysis shows.[1] Young people in particular are placing higher demands on themselves and on others. Our dog-eat-dog world, full of impeccable images of what our bodies, careers and homes should look like, seems to be creating a rising tide of people who may be putting themselves at risk of mental and physical illness in their search for the perfect life. It all sounds quite alarming, but there are solutions: ways to learn when good is good enough.

It is a little hard to define perfectionism. Most psychological studies, however, measure it using a standard scale that consists of forty-five statements, such as 'I strive to be the best at everything I do', or 'If I ask someone to do something, I expect it to be done flawlessly'. You rate how much you agree with each, and if you identify with a lot of them, it's likely you have perfectionist tendencies.

Psychologists also distinguish between three different kinds of perfectionism. 'Self-oriented' perfectionists set themselves high goals in their work and relationships. They can often experience anxiety from losing to a competitor, failing at a test or not getting a bonus at work. 'Other-oriented' perfectionists hold those around them to exceptionally high standards. They are very critical and judgemental of others and risk social rejection and relationship problems. Finally, there are 'socially prescribed' perfectionists, who feel immense pressure from others to be perfect while also seeking their approval. The impossibly high standards this later group set for themselves mean they often feel harshly scrutinised or rejected. Their self-esteem can take a hit on a daily basis and they feel a lot of negative emotions as a result.

But where does the line between being a perfectionist and just having high standards lie? Often, it comes down to how a person deals with success or failure. If someone is hard-working and diligent, they will tend to appreciate any success, and adapt their goals when they fail. They can put in just enough effort for a strategic benefit. A perfectionist will take much less pleasure from positive outcomes. Even a notable success, or a goal achieved, will be met with a sense of pressure that they need to keep up this level of success, or that the goalposts need to be placed further away.

Although perfectionism does have a genetic element to it – people are more likely to show these tendencies if their parents also have them – the environmental factors that push a person towards perfectionism have increased in recent decades, says Thomas Curran at the University of Bath in the UK. He says that a generation ago in a country such as the UK, the government took more responsibility for education and there was less pressure to achieve. Now, young people take on their own risk for success and failure. They pay for university, there is more standardised testing from a younger age, more competition over good schools, alongside having to deal with social media setting unrealistic targets for other aspects of life. Throw a dodgy economy into the mix and you have an unprecedented storm of pressure to reach unattainable targets.

You might not think perfectionism sounds like the worst habit to have. And you'd be right in thinking that perfectionists do tend to achieve highly academically, or in their career. But it has a definite dark side. Although perfectionism isn't considered a syndrome in its own right, it has strong links with mental health conditions that are.[2] People with eating disorders, OCD, and depression have higher levels of perfectionism compared with people without these conditions. Perfectionism has also been linked with stress and an increased risk of cardiovascular disease;[3] and those with the trait often experience higher levels of anger and anxiety and are more likely to commit suicide.

Perfectionism becomes more problematic as we age.[4] It sounds counter-intuitive, but perfectionists become less conscientious, less diligent, less productive, increasingly neurotic and more likely to experience burnout. It makes sense when you think about it – if you strive for perfection, you're likely to perceive a high frequency of failures and low frequency of successes. Over time, you become increasingly disengaged from work and emotionally fraught. Rather than opening yourself up to failure, you shut down and don't try in the first place. I have a friend whose perfectionism eventually led to a nervous breakdown, ultimately triggered by not being able to decide what outfit to wear to her job in the morning. She had cultivated an impeccable look and just couldn't reach her own standards of perfection. She was paralysed by the indecision and eventually needed medical intervention.

So what can we do to combat perfectionism? The first thing is to recognise it, although by the very nature of the condition it is extremely hard for a perfectionist to admit there might be something wrong. Talking therapies or cognitive behavioural therapy (CBT), which give people tools to analyse why they are acting in certain ways and enable them to break out of negative patterns, are the best way to combat the behaviour.

If you're not able to access these, you might find that forcing yourself to step out of your comfort zone helps. From the experiences witnessed by therapists, pushing yourself to do things when you know the conditions aren't perfect – from working out when you're not feeling one hundred per cent, or stopping after a given time rather than when you think your project is perfect – is a good start that can make a big difference over time.

The final, essential piece of advice I gained from talking to experts who study perfectionism is this: while 'perfect' is a non-existent state, learning and improving yourself is a worthy life-long aspiration. If you can promote life-long learning as an attitude in your children and as a mantra in your own life, not only will this boost that vital cognitive reserve that will protect your mental and physical faculties well into old age, but it will also ensure you maintain the sense that things are not over when one test is done, one race has been won or lost, one hurdle has been overcome. Horrible old cliché it may be, but when it comes to fixing your life it's not the destination, but what you do on your way there, that counts.

AFTERWORD

I am conscious that using serious, evidence-based studies is not, by any means, the easiest way to go about fixing your life. It is much simpler to ask a friend what works for them, listen to old wives' tales, perhaps just do what your mum did before you. But if you decide to take the easier path, you must also accept that it might be a complete waste of time.

Science truly is the best way to understand what works in this world and what doesn't. If it hasn't been tested, then it is as anecdotal as the personal stories that I have interspersed throughout these chapters. That's not to say that what your mum or your friends or the old wives' tales say are always wrong. Maybe their advice is good, maybe it isn't. But with only limited time in the day to make changes that have the potential to add entire years of better physical and mental health to our life, I know which I'd rather put my money on.

That said, I don't want you to think it's possible to walk around living your life by scientific studies 24/7. That's just not feasible, nor enjoyable. I certainly don't. I don't put shoes on tables or open umbrellas indoors because my mum once told me it was bad luck. Tomorrow, you'll find me baking 'labour cookies' because two of my friends swore by their ability to bring on the birth of their children. I have a lucky pair of underpants, for goodness sake.

What I'm trying to say is that there's nothing wrong with

occasionally listening to friends, to instinct, to a person you admire from the telly; perhaps some of the things they do will work for you too. Maybe it's just fun to try something new. Sometimes these things will be life-changing. Sometimes things that have been tested on millions of people won't work for you – someone has to be an outlier. Life is short, so whatever path you choose to follow, all I ask is that you do so with better knowledge of what you're letting yourself in for.

You'll have noticed that I chose to summarise five take-home points at the end of the chapters. This wasn't very scientific of me. These haven't been approved by scientists, or selected for their special power to change your life, but because I felt they might offer a few quick, evidence-based tips for those of you who are particularly time-poor. But I urge you to find the time to read in full the chapters that are most important to you, to understand how to avoid bad advice, make yourself more aware of your unconscious biases, work out how to ask the right questions when you see extraordinary headlines, and take the time to really get to grips with your own behaviours, so that you can make the most of your future.

And, finally, don't stop here. Science is a work in progress, not the final answer. Good self-help will always need reshaping, re-analysing, and refreshing periodically to make sure it really is the best recommendation. So find some lucky pants or follow this guide to the letter – either way, make sure you're constantly on the lookout for even better advice. You never know, it might just be the fix that changes your life.

ACKNOWLEDGEMENTS

This book was not a solo project. First, all of the information, data, research, quotes, knowledge and statistics came from the minds of thousands of dedicated scientists over several decades. Second, almost everything you have read was first conceived, conjugated and perfected by a team of staff and freelance writers and editors at *New Scientist*.

I have combined my own research, stories and knowledge with features and news from the magazine's archive, and I would be remiss not to individually acknowledge each and every one of these wonderful journalists whose work I have incorporated into the book. I apologise unreservedly if I have missed anyone off this list. Thank you William Lee Adams, Sally Adee, Pragya Agarwal, Anil Ananthaswamy, Alun Anderson, Abigail Beall, Jessica Bond, Michael Bond, Catherine Brahic, Teal Burrell, Michelle Carr, Eleanor Case, Catherine de Lange, Kate Douglas, Robin Dunbar, Madeleine Finlay, Linda Geddes, Alison George, Jessica Hamzelou, Martie Haselton, Douglas Heaven, Rowan Hooper, Joshua Howgego, Christian Jarrett, Dan Jones, Chloe Lambert, Graham Lawton, Jo Marchant, Alison Motluk, Tiffany O'Callaghan, Heather Pringle, David Raubenheimer, Timothy Revell, David Robson, Moya Sarner, Megan Scudellari, Chris Simms, Stephen Simpson, Chris Stokel-Walker, Kayt Sukel, Amelia Tait, Sonia Van Gilder Cooke, Susan Watts, Jon White, Caroline Williams, Sam Wong and Emma Young.

A massive thank you to Richard Webb who stepped in when imminent labour put pause to my finishing touches, and at John Murray Press to Georgina Laycock and Abigail Scruby for their wonderful editing and ideas, and Yassine Belkacemi for his consistently amazing work in promoting the books that I write.

Finally, thank you to Alex, Jess and baby Sam for being the most wonderful family in the world.

NOTES

Chapter 1: How Not to Worry

1. Morgan, C. A. et al. (2002), 'Neuropeptide-Y, Cortisol, and Subjective Distress in Humans Exposed to Acute Stress: Replication and Extension of Previous Report', *Biological Psychiatry* 52, 2, 136–42.
2. Chao, L. L. et al. (2015), 'Preliminary Evidence of Increased Hippocampal Myelin Content in Veterans with Posttraumatic Stress Disorder', *Frontiers in Behavioural Neuroscience* 9, 333.
3. Johnson, D. C. et al. (2014), 'Modifying Resilience Mechanisms in At-Risk Individuals: A Controlled Study of Mindfulness Training in Marines Preparing for Deployment', *American Journal of Psychiatry* 171, 8, 844–53.
4. De With, M. et al. (2019), 'Effects of Music Interventions on Stress-Related Outcomes: A Systematic Review and Two Meta-Analyses', *Health Psychology Review* 14, 2, 1–31.
5. Mujcic, R. and Oswald, A. J. (2016), 'Evolution of Well-Being and Happiness After Increases in Consumption of Fruit and Vegetables', *American Journal of Public Health* 106, 8, 1504–10.
6. Everitt, H. et al. (2019), 'Therapist Telephone-Delivered CBT and Web Based CBT Compared With Treatment As Usual in Refractory Irritable Bowel Syndrome: The ACTIB Three-Arm RCT', *Health Technology Assessment* 23, 17, 1–154.
7. Stonerock, G. L. et al. (2015), 'Exercise as Treatment for Anxiety: Systematic Review and Analysis', *Annals of Behavioural Medicine* 49, 4, 542–56.
8. Schmidt, K. et al. (2014), 'Prebiotic Intake Reduces the Waking Cortisol Response and Alters Emotional Bias in Healthy Volunteers', *Psychopharmacology* 232, 10, 1793–1801.

Chapter 2: How to be Happy

1. Gilbert, D. T. and Ebert, J. E. J. (2002), 'Decisions and Revisions: The Affective Forecasting of Changeable Outcomes', *Journal of Personality and Social Psychology* 82, 4, 503–14.

2. Tang, Y.-Y. et al. (2015), 'The Neuroscience of Mindful Meditation', *Nature Reviews Neuroscience* 16, 4, 213–25.
3. Fredrickson, B. L. et al. 'Open Hearts Build Lives: Positive Emotions, Induced Through Loving-Kindness Meditation, Build Consequential Personal Resources', *Personality and Social Psychology* 95, 5, 1045–62.
4. Joanna, F. W. et al. (2006), 'The Effect of Positive Writing on Emotional Intelligence and Life Satisfaction', *Journal of Clinical Psychology* 62, 10, 1291–302.
5. Mitchell, R. and Popham, F. (2008), 'Effect of Exposure to Natural Environment on Health Inequalities: An Observational Population Study', *Lancet* 372, 9650, 1655–60.
6. Aerts, R. et al. (2018), 'Biodiversity and Human Health: Mechanisms and Evidence of the Positive Health Effects of Diversity in Nature and Green Spaces', *British Medical Bulletin* 127, 1, 5–22.
7. Saxbe, D. E. and Repetti, R. (2009), 'No Place Like Home: Home Tours Correlate with Daily Patterns of Mood and Cortisol', *Personality and Social Psychology* 36, 1, 71–81.
8. Bharwani, A. et al. (2017), 'Oral Treatment with *Lactobacillus rhamnosus* Attenuates Behavioural Deficits and Immune Changes in Chronic Social Stress', *BMC Medicine* 15, 7.
9. Strandwitz, P. et al. (2019), 'GABA-Modulating Bacteria of the Human Gut Microbiota', *Nature Microbiology* 4, 396–403.

Chapter 3: How to be More Confident

1. Ott, T. et al. (2019), 'The Neurobiology of Confidence: From Beliefs to Neurons', *Cold Spring Harbour Symposia on Quantitate Biology* 83, 9–16.
2. Ibid.
3. Go Johnny 5!
4. Van Dyck, E. (2019), 'Musical Intensity Applied in the Sports and Exercise Domain: An Effective Strategy to Boost Performance?', *Frontiers in Psychology* 10, 1145.
5. Lammers, J. et al. (2013), 'Power Gets the Job: Priming Power Improves Interview Outcomes', *Journal of Experimental Social Psychology* 49, 4, 776–9.
6. Carney, D. R. et al. (2010), 'Power Posing: Brief Nonverbal Displays Affect Neuroendocrine Levels and Risk Tolerance', *Psychological Science* 21, 10, 1363–8.
7. Simmons, J. P. and Simonsohn, U. (2017), 'Power Posing: *P*-curving the Evidence', *Psychological Science* 28, 5, 687–93.

8. Cuddy, A. J. C. et al. (2017), 'P-curving a More Comprehensive Body of Research on Postural Feedback Reveals Clear Evidential Value for Power-Posing Effects: Reply to Simmons and Simonsohn', *Psychological Science* 29, 4, 656–66.

9. Havas, D. A. et al. (2010), 'Cosmetic Use of Botulinum Toxin-A Affects Processing of Emotional Language', *Psychological Science* 21, 7, 895–900.

10. Dunning, D. and Ehrlinger, J. (2003), 'How Chronic Self-Views Influence (and Potentially Mislead) Estimates of Performance', *Journal of Personality and Social Psychology* 84, 1, 5–17.

11. Late, I. M. et al. (2013), 'Successful Female Leaders Empower Women's Behavior in Leadership Tasks', *Journal of Experimental Social Psychology* 49, 444–8.

12. Nettle, D. (2005), 'An Evolutionary Approach to the Extraversion Continuum', *Evolution and Human Behavior* 26, 4, 363–73.

13. Wald, I. et al. (2013), 'Attention to Threats and Combat-Related Posttraumatic Stress Symptoms', *JAMA Psychiatry* 70, 4, 401–8.

Chapter 4: How to Make Friends

1. Meshi, D. et al. (2013), 'Nucleus Accumbens Response to Gains in Reputation for the Self Relative to Gains for Others Predicts Social Media Use', *Frontiers in Human Neuroscience* 7, 439.

2. Fowler, J. H. et al. (2009), 'Model of Genetic Variation in Human Social Networks', *PNAS* 106, 6, 1720–4.

3. Christakis, N. A. and Fowler, J. H. (2004), 'Friendship and Natural Selection', *PNAS* 111, 3, 10796–801.

Chapter 5: How to Find Love

1. Acevedo, B. P. et al. (2012), 'Neural Correlates of Long-Term Intense Romantic Love', *Social Cognitive and Affective Neuroscience* 7, 2, 145–59.

2. Bruch, E. E. and Newman, M. E. J. (2018), 'Aspirational Pursuit of Mates in Online Dating Markets', *Science Advances* 4, 8.

3. Vallejo, A. G. et al. (2019), 'Propofol-Induced Deep Sedation Reduces Emotional Episodic Memory Reconsolidation in Humans', *Science Advances* 5, 3.

4. DeWall, C. N. et al. (2010), 'Acetaminophen Reduces Social Pain: Behavioural and Neural Evidence', *Psychological Science* 21, 7, 931–7.

5. https://yougov.co.uk/topics/lifestyle/articles-reports/2015/05/27/one-five-british-adults-admit-affair

Chapter 6: How to Live Healthier for Longer

1. Tang, D. W. et al. (2014), 'Behavioural and Neural Valuation of Foods is Driven by Implicit Knowledge of Caloric Content', *Psychological Science* 12, 2168–76.
2. Buman, M. P. et al. (2008), 'Hitting the Wall in the Marathon: Phenomenological Characteristics and Associations with Expectancy, Gender, and Running History', *Psychology of Sport and Exercise* 9, 177–90.
3. Stevenson, C. D. and Biddle, S. J. (1998), 'Cognitive Orientations in Marathon Running and "Hitting the Wall"', *British Journal of Sports Medicine* 32, 229–35.

Chapter 7: How to Sleep Well

1. Simon, C. W. and Emmons, W. H. (1956), 'Responses to Material Presented During Various Levels of Sleep', *Journal of Experimental Psychology* 51, 89–97.
2. Arzi, A., et al. (2015), 'Olfactory Aversive Conditioning During Sleep Reduces Cigarette-Smoking Behavior', *Journal of Neuroscience* 34, 46, 15382–93.

Chapter 8: How to Make and Break Habits

1. Neal, D. T. et al. (2011), 'The Pull of the Past: When Do Habits Persist Despite Conflict With Motives?', *Personality and Social Psychology Bulletin* 37, 11, 1428–37.
2. Lally, P. et al. (2009), 'How are Habits Formed: Modelling Habit Formation in the Real World', *European Journal of Social Psychology* 40, 6, 998–1009.
3. Caspi, A. et al. (2017), 'Childhood Forecasting of a Small Segment of the Population With Large Economic Burden', *Nature Human Behaviour* 1, 0005.
4. Job, V. et al. (2010), 'Ego Depletion – Is it All in Your Head?: Implicit Theories About Willpower Affect Self-Regulation', *Psychological Science* 21, 11, 1686–93.
5. Miller, E. M. et al. (2012), 'Theories of Willpower Affect Sustained Learning', *PLOS ONE* 7, 6, e38680.
6. Bernecker, K. et al. (2015), 'Implicit Theories About Willpower Predict Subjective Well-Being', *Journal of Personality* 85, 2, 136–50.
7. Lindson, N. et al. (2019), 'Smoking Reduction Interventions for Smoking Cessation', *Cochrane Database Systematic Reviews* 9, 1–190.
8. George, J. et al. (2019), 'Cardiovascular Effects of Switching from Tobacco Cigarettes to Electronic Cigarettes', *Journal of the American College of Cardiology* 74, 25, 3112–20.

9. Hajek, P. et al. (2019), 'A Randomized Trial of E-Cigarettes Versus Nicotine-Replacement Therapy', *New England Journal of Medicine* 380, 7, 629–37.
10. Frings, D. et al. (2020), 'Comparison of Allen Carr's Easy Way Programme With a Specialist Behavioural and Pharmacological Smoking Cessation Support Service: A Randomized Controlled Trial', *Addiction* 115, 5, 977–85.
11. Ma, Y. et al. (2015), 'The Significant Association of Taq1A Genotypes in DRD2/ANKK1 with Smoking Cessation in a Large-Scale Meta-Analysis of Caucasian Populations', *Translational Psychiatry* 5, e686.
12. Johnson, K. E. and Voight, B. F. (2018), 'Patterns of Shared Signatures of Recent Positive Election Across Human Populations', *Nature Ecology and Evolution* 2, 4, 713–20.

Chapter 9: How to be Smarter

1. Deary, I. J. et al. (2007), 'Intelligence and Educational Achievement', *Intelligence* 35, 1, 13–21.
2. Forrest, L. F. et al. (2011), 'The Influence of Childhood IQ and Education on Social Mobility in the Newcastle Thousand Families Birth Cohort', *BMC Public Health* 11, 895.
3. Flynn, J. R. (2007), *What Is Intelligence? Beyond the Flynn Effect*. Cambridge: Cambridge University Press.
4. Ritchie, S. J and Tucker-Drop, E. M. (2018), 'How Much Does Education Improve Intelligence? A Meta-Analysis', *Psychological Science* 29, 8, 1358–69.
5. Smart, E. L. et al. (2014), 'Occupational Complexity and Lifetime Cognitive Abilities', *Neurology* 83, 24.
6. Rauscher, F. H. et al. (1993), 'Music and Spatial Task Performance', *Nature* 365, 611.
7. Hampshire, A. et al. (2019), 'A Large-Scale, Cross-Sectional Investigation Into the Efficacy of Brain Training', *Frontiers in Human Neuroscience* 13, 221.
8. Greely, H. et al. (2008), 'Towards Responsible Use of Cognitive-Enhancing Drugs by the Healthy', *Nature* 456, 702–5.
9. Cowan, N. (2010), 'The Magical Mystery Four: How is Working Memory Capacity Limited, and Why?', *Current Direct Psychological Science* 19, 1, 51–7.
10. Wang, Q. (2013), 'Gender and Emotion in Everyday Event Memory', *Memory* 21, 4, 503–11.
11. Persson, J. et al. (2013), 'Remembering Our Origin: Gender Differences in Spatial Memory Are Reflected in Gender Differences in Hippocampal Lateralization', *Behavioural Brain Research* 256, 219–28.

12. Dresler, M. et al. (2017), 'Mnemonic Training Reshapes Brain Networks to Support Superior Memory', *Neuron* 93, 5, 1227–35.

13. Mangen, A. et al. (2019), 'Comparing Comprehension of a Long Text Read in Print Book and on Kindle: Where in the Text and When in the Story?', *Frontiers in Psychology* 10, 38.

14. Sparrow, B. et al. (2011), 'Google Effects on Memory: Cognitive Consequences of Having Information at Our Fingertips', *Science* 333, 6043, 776–8.

15. Storm, B. C. and Stone, S. M. (2015), 'Saving-Enhanced Memory: The Benefits of Saving on the Learning and Remembering of New Information', *Psychological Science* 26, 2, 182–8.

16. Karpicke, J. D. and Roediger, H. L. (2008), 'The Critical Importance of Retrieval for Learning', *Science* 319, 966–8.

17. Pashler, H. et al. (2007), 'Enhancing Learning and Retarding Forgetting: Choices and Consequences', *Psychonomic Bulletin & Review* 14, 187–93.

18. Sommerlad, A. et al. (2018), 'Marriage and Risk of Dementia: Systematic Review and Meta-Analysis of Observational Studies', *Journal of Neurology, Neurosurgery & Psychiatry* 89, 231–8.

Chapter 10: How to be Wiser

1. Grossmann, I. et al. (2013), 'A Route to Well-Being: Intelligence Versus Wise Reasoning', *Journal of Experimental Psychology: General* 142, 3, 944–53.

2. Hafenbrack, A. C. et al. (2013), 'Debiasing the Mind Through Meditation: Mindfulness and the Sunk-Cost Bias, *Psychological Science* 25, 2, 369–76.

3. Vazire, S. and Carlson, E. N. (2011), 'Others Sometimes Know Us Better Than We Know Ourselves', *Current Direction in Psychological Science* 20, 2, 104–8.

4. Anderson, C. et al. (2012), 'A Status-Enhancement Account of Overconfidence', *Journal of Personality and Social Psychology* 103, 4, 718–35.

5. Zschirnt, E. and Ruedin, D. (2016), 'Ethnic Discrimination in Hiring Decisions: A Meta-Analysis of Correspondence Tests 1990–2015', *Journal of Ethnic and Migration Studies* 42, 7, 1115–34.

6. Goldin, C. and Rouse, C. (2000), 'Orchestrating Impartiality: The Impact of "Blind" Auditions on Female Musicians', *American Economic Review* 90, 4, 715–41.

7. Rinne, U. (2018), 'Anonymous Job Applications and Hiring Discrimination', *IZA World of Labor*. https://doi.org/10.15185/izawol.48.v2

8. Drwecki, B. B., Moore, C. F., Ward, S. E. and Prkachin, K. M. (2011), 'Reducing Racial Disparities in Pain Treatment: The Role of Empathy and Perspective-Taking', *Pain* 152, 5, 1001–06.

Chapter 11: How to Get Ahead in Life

1. Ruffle, B. J. and Shtudiner, Z. (2015), 'Are Good-Looking People More Employable?', *Management Science* 61, 8, 4–6.
2. Howlett, N. et al. (2015), 'Unbuttoned: The Interaction Between Provocativeness of Female Work Attire and Occupational Status', *Sex Roles: A Journal of Research* 72, 3–4, 105–16.
3. Danziger, S. et al. (2011), 'Extraneous Factors in Judicial Decisions', *PNAS* 108, 17, 6889–92.
4. Tormala, Z. L. and Petty, R. E. (2002), 'What Doesn't Kill Me Makes Me Stronger: The Effects of Resisting Persuasion on Attitude Certainty', *Journal of Personality and Social Psychology* 83, 6, 1298–313.
5. Zimmerman, D. H. and West, C. (1975), 'Sex Roles, Interruptions and Silences in Conversation', *Language and Sex: Difference and Dominance*, 105–29.
6. Jacobi, T. and Schweers, D. (2017), 'Justice, Interrupted: The Effect of Gender, Ideology and Seniority at Supreme Court Oral Arguments', *Virginia Law Review* 103, 7, 1379–496.
7. Stivers, T. et al. (2009), 'Universals and Cultural Variation in Turn-Taking in Conversation', *PNAS* 106, 26, 10587–92.
8. Han, Z. L. (2001), 'Cooperative and Intrusive Interruptions in Inter- and Intracultural Dyadic Discourse', *Journal of Language and Social Psychology* 20, 3, 259–84.
9. Kennedy, C. and Camden, C. (1983), 'Interruptions and Nonverbal Gender Differences', *Journal of Nonverbal Behaviour* 8, 91–108.
10. Tuckman, B. W. (2005), 'Relations of Academic Procrastination, Rationalizations, and Performance in a Web Course with Deadlines', *Psychological Reports* 96, 1015–21.
11. Sirois, F. M. (2007). '"I'll Look After My Health, Later": A Replication and Extension of the Procrastination Model With Community-Dwelling Adults', *Personality and Individual Differences* 43, 1, 15–26.
12. Bellezza, S. et al. (2017), 'Conspicuous Consumption of Time: When Busyness and Lack of Leisure Time Become a Status Symbol', *Journal of Consumer Research* 44, 1, 118–38.
13. Dunstan, D. W. et al. (2012), 'Too Much Sitting – Health Hazard', *Diabetes and Clinical Practice* 97, 3, 368–76.
14. Golder, S. A. and Macy, M. W. (2011), 'Diurnal and Seasonal Mood Vary with Work, Sleep, and Day-Length Across Diverse Cultures', *Science* 333, 6051, 1878–881.
15. Kühnel, J. et al. (2017), 'Take a Break! Benefits of Sleep and Short Breaks for Daily Work Engagement', *European Journal of Work and Organisational Psychology* 26, 4, 481–91.

16. Lamote de Grignon Pérez, J. et al. (2019), 'Sleep Differences in the UK Between 1974 and 2015: Insights From Detailed Time Diaries', *Journal of Sleep Research* 28, 1, e12753.

17. Self, P. et al. (2016), 'Mind-Wandering With and Without Intention', *Trends in Cognitive Sciences* 20, 8, 605–17.

18. Self, P. et al. (2015), 'Not All Mind Wandering is Created Equal: Dissociating Deliberate From Spontaneous Mind Wandering', *Psychological Research* 79, 750–8.

19. Self, P. et al. (2015), 'On the Relation of Mind Wandering and ADHD Symptomatology', *Psychonomic Bulletin & Review* 22, 629–36.

20. Smallwood, J. and Andrews-Hanna, J. (2013), 'Not All Minds That Wander Are Lost: The Importance of a Balanced Perspective on the Mind-Wandering State', *Frontiers in Psychology* 16, 4, 441.

21. Ibid.

22. Self, P. et al. (2016), 'Mind-Wandering With and Without Intention', *Trends in Cognitive Sciences* 20, 8, 605–17.

23. Cousins, J. N. et al. (2019), 'The Long-Term Memory Benefits of a Daytime Nap Compared With Cramming', *Sleep* 42, 1.

24. Jarosz, A. F. et al. (2012), 'Uncorking the Muse: Alcohol Intoxication Facilitates Creative Problem Solving', *Consciousness and Cognition* 21, 1, 487–93.

Chapter 12: How to Avoid Being Too Perfect

1. Curran, T. and Hill, A. P. (2019), 'Perfectionism is Increasing Over Time: A Meta-Analysis of Birth Cohort Differences From 1989 to 2016', *American Psychological Association* 145, 4, 410–29.

2. Limburg, K. et al. (2016), 'The Relationship Between Perfectionism and Psychopathology: A Meta-Analysis', *Journal of Clinical Psychology* 73, 10, 1301–26.

3. Corson, A. T. et al. (2018), 'Perfectionism in Relation to Stress and Cardiovascular Disease Among Gifted Individuals and the Need for Affective Interventions', *Roeper Review* 40, 1, 46–55.

4. Smith, M. M. et al. (2019), 'Perfectionism and the Five-Factor Model of Personality: A Meta-Analytic Review', *Personality and Social Psychology Review* 23, 4, 367–90.

INDEX

317

INDEX

cinema dates 99
circadian rhythms
  chronotypes 280–1
  emotions and energy levels 280
  light exposure 40
  sleep 155–6, 159, 199
Clark, Brian 138
Clockwork Orange, A (Burgess) 160
clothing
  confidence 58
  dating 99–100
  interviews 267
clutter 41–2
coffee 135, 149, 159, 281
cognition
  embodied 31, 61–2
  enclothed 58
cognition-enhancing drugs 213–15
cognitive behavioural therapy (CBT)
    25–6, 36, 86–7, 303
cognitive bias 235, 236–40
cognitive capacity: effect of happiness 38
cognitive interference 58
cognitive reserve 225–6, 303
combat survival 263
comedy 15
communication
  interruption prevention 273–6
  long-term partners 105
  persuasion 270–2
  unconscious language skills 250
  see also conversation
compassionate meditation 36
competition
  education 302
  uncertainty and confidence levels 67
compulsion *see* addiction
concentration 129, 149, 279–86
Concordia Longitudinal Risk Project 243
confidence
  body language 58–62
  clothing 58
  dating 100
  gender differences 62–5
  interview training 51–2
  introversion 65–8
  music 56–7, 68
  points of view 271–2
  procrastination effect 277
  self-worth 241
  subjectivity 53
  top tips 68
conflict avoidance 256
Conley, Terri 112

conscientiousness 55
contamination experiment 19
contempt in relationships 106
contraception 102, 112
conversation
  dating 100
  interruption 274–5
  small talk 77–9
  see also communication
Cookie Monster 173
cooperation 74, 85, 101
Cornell, Ed 290–1
coronavirus pandemic 9–10, 17, 18, 299
cortex 11
cortisol
  health risks 14
  home tidiness 41–2
  microbiome 45
  perceived fitness levels 137
  stress response 11, 12–13, 139
cravings
  environmental cues 171–2
  exercise effects 127
  microbiome 131, 133–4
  relationship break-ups 107
  reward system 131–2
  willpower 173
creativity
  alcohol 225, 288–9
  exercise effects 127, 130, 287
  flow 289
  increasing 286–9
  mind-wandering 284, 287–8
  REM sleep 150–1
  time of day 280
critical thinking 235
criticism in relationships 106
Cuddy, Amy 59–61
cultural attitudes
  interruption 274–5
  stereotypes 244–7
  willpower 176
curiosity 17
Curran, Thomas 301

dairy products 44, 45, 157
Dalai Lama 35
dance 73
Dante Alighieri 17
Darwin, Charles 93, 104
dating
  film choices 99
  first date impressions 97–8
  online 96–7, 101
  speed-dating 78, 79

INDEX